CREATION

CREATION
LIFE AND HOW TO MAKE IT

* *

STEVE GRAND

Harvard University Press
Cambridge, Massachusetts
2001

Printed in the United States of America

First United Kingdom Publication 2000
by Weidenfeld & Nicolson

Library of Congress Cataloging-in-Publication Data

Grand, Steve.
Creation : life and how to make it / Steve Grand.
p. cm.
Includes bibliographical references (p.).
ISBN 0-674-00654-2 (alk. paper)
1. Biological systems—Computer simulation.
2. Artificial intelligence. 3. Artificial life. I. Title
QH324.2 .G73 2001
570'.1'13—dc21   2001024165

To Mum and Dad for my past
To Ann for my present
To Chris for his future

CONTENTS

* *

ACKNOWLEDGEMENTS

* *

Behind every individual event in this universe lies a complex web of cause and effect that stretches back into infinity. Here are a few of the causative agents that I would like to single out and thank for making this particular literary event possible.

First, thanks to Peter Tallack at Weidenfeld & Nicolson for having the good sense to commission this book, the skill to offer me sound advice and the patience to pretend not to mind when I whinged about having to take heed of it. Also thanks to John Woodruff for dotting all the t's and crossing all the i's (you might want to check that bit, John) and making the final copy readable. I am grateful to Richard Dawkins on several counts: for his own writings; for treating my crazy ideas with undue respect; for telling me that studying for a PhD would be a waste of my time and I would do better to write a book; and finally for setting in train the events that led to the opportunity to do so. Stretching further back into the causal mists, my thanks go to Michael Hayward for having the faith to let me write *Creatures*, and the patience to stick with it when a simple nine-month project blossomed into a five-year personal quest. Thanks also to Creature Labs for permission to write about the project, and of course to my fellow programmers and artists for helping to turn an idea into a real product. I am grateful to Dave Cliff for bringing me in from the wilderness and introducing me to the scientific community, and for endorsing *Creatures* as a worthwhile piece of science. Two sources of general inspiration especially deserve citing: the first is Douglas Adams for putting into dramatic form concepts that I have only ever been able to describe to a computer; and the second is A.K. Dewdney, whom I have never met, for writing a slim but inspiring work of fiction entitled *The Planiverse*, which focused my desire to attempt the creation of artificial life. Finally, I would like to thank my parents for teaching me to think for myself, and of course my wife Ann and my son Christopher for putting up with all the angst and adding a spin of their own to the ideas that I bounced off them. Anyone else I should have mentioned will probably know what an awful memory I have and will forgive me.

Perhaps this is also the place to apologize to those scientists and philosophers whose work I have not acknowledged. Throughout this book you will find the phrase 'artificial life' used to describe a synthetic living being. There is also an established scientific discipline with the same name. These two are not quite synonymous, and so to differentiate them I shall be using the term 'A-life' when I'm referring to the discipline. Many brilliant people have done sterling work in this and related fields of enquiry, and we shall meet one or two of them in these pages. Yet A-life is not the same thing as the quest to create artificial life. It concerns itself with the broader topic of studying lifelike phenomena through the use of computer simulation, and only rarely are these phenomena assembled together into complete functioning organisms, and that is the topic of this book. Like many of the people researching what is now called A-life, I spent a great deal of my time in the wilderness, working in isolation, and only fairly recently did I discover the degree to which others shared (and frequently preceded) my own thinking. This is largely a personal book, and I hope none of my colleagues feel slighted if I fail to describe their work or omit to credit their discoveries. This is not meant to be a historical introduction to the fields of A-life and artificial intelligence (AI), nor do I claim to be relating the 'establishment view' of either of these topics. For those who are interested in the broader background, I have listed some useful books in the Bibliography. Rather than being an overview of the A-life field, you might look upon this book as a 'Zen' guide to how to be a digital god.

Steve Grand
Shipham, Somerset, June 2000

* *

A LATTER-DAY FRANKENSTEIN

What a queer thing Life is! So unlike anything else, don't you know, if
you see what I mean.

P.G. Wodehouse, *My Man Jeeves*

As the words for this introduction start to form in my mind, I am lying
in bed, curled around my wife's sleeping body. As I hold her I enjoy the
sensation of her presence – her touch, her scent, the ebb and flow of
her breathing. Yet it is clearly not the sensations themselves that I
savour. If someone were to replace her with a cunningly designed
machine, warmed to 36 degrees Celsius, suffused with synthetic
pheromones, programmed to wriggle engagingly and to emit a gentle
rhythmic hissing noise, it would do nothing for me at all. It is not the
sensations but their meaning that keeps me awake and captivated. To
put it in a rather more macabre way: suppose Ann died right now, here
in my arms. I imagine I would continue to hold her awhile, but it
would no longer really be her that I'd be holding, just her memory. Her
inner being, her *élan vital*, would have gone. This is a book about that
vital essence – a scientific search for the soul. More than that, more
adventurously, more arrogantly, more grandly than that, it's a book
that tries to answer the question, 'How can we *build* a soul?'

The vital spark

One of the phenomena which had peculiarly attracted my attention was
the structure of the human frame, and, indeed, any animal endued with
life. Whence, I often asked myself, did the principle of life proceed? It
was a bold question, and one which has ever been considered as a

mystery; yet with how many things are we upon the brink of becoming acquainted, if cowardice or carelessness did not restrain our enquiries. I resolved these circumstances in my mind, and determined thenceforth to apply myself more particularly to those branches of natural philosophy which relate to physiology ...

After days and nights of incredible labour and fatigue, I succeeded in discovering the cause of generation and life; nay, more, I became myself capable of bestowing animation upon lifeless matter ...

It was on a dreary night of November that I beheld the accomplishment of my toils. With an anxiety that almost amounted to agony, I collected the instruments of life around me, that I might infuse a spark of being into the lifeless thing that lay at my feet. It was already one in the morning; the rain pattered dismally against the panes, and my candle was nearly burnt out, when, by the glimmer of the half-extinguished light, I saw the dull yellow eye of the creature open; it breathed hard, and a convulsive motion agitated its limbs.

Mary Shelley, *Frankenstein*

When Victor Frankenstein breathed life into his creation, it was with electricity. He, or rather his own creator, Mary Shelley, refused to allow us access to the exact recipe, but it seems that Frankenstein constructed his monster laboriously from ordinary, everyday chemicals and then finally animated him by means of a vital spark.

Electricity was the miracle of Mary Shelley's time. A generation earlier, in 1780, Luigi Galvani had demonstrated the significance of electricity in the processes of life, by showing how amputated frogs' legs would twitch when momentarily connected to a crude electric battery made from body fluids and two dissimilar metals. If the lifeless limb of a frog could be made to behave as if alive by the infusion of the electric fluid, then surely electricity itself must be the source and very essence of life?

The idea that living things possess some special substance that separates them from inanimate matter and gives them their astonishing properties goes back a long way. When they weren't occupied with turning base metals into gold, the ancient alchemists were intent on discovering the *elixir vitae*, or water of life, which they hoped would enable them to cheat death and live for ever. This was certainly their technological aim, as it were, but they also had what today we would call a pure-science motive for their work: they wished to understand life; they wanted to divine the nature of the soul.

We human beings are obsessed with our souls. Painters don't paint portraits because they enjoy applying pigment to canvas. Poets don't write for the joy of seeing squiggles on paper. Both want to capture some essence, some aspect of that which we call the human condition. What happens to us when we die is, of course, what strikes us as the most pertinent and sometimes the most pressing question when we dwell upon the topic of soul. One day, each of us will face death. Ideally, most of us would prefer to forego the experience altogether, but at the very least we would like some hints, if not assurances, as to what will befall us when our time comes. But before we die we are alive, and life itself is such a strange phenomenon, something so distinctive and inexplicable, that I think we are all deeply fascinated by what it means. What is life? How does it come about? What exactly *is* the 'answer to the ultimate question of life, the universe and everything'?

The notion that life resides in some special substance, which is excused from compliance with the normal laws of physics, is called vitalism. In the late Middle Ages this vital essence was deemed to be chemical in nature. By the early nineteenth century it had become electrical. As our mastery over the chemical and electrical world increased, and we found ourselves still no nearer to discovering the essence of life, the concept was pushed into even more abstract realms. The vague notion of a 'life force' surfaces today in science fiction, with the implication that life is akin to magnetism or gravity. Vitalism is not yet dead by any means, but the vital essence continues to flee into whatever phenomenal realm currently lies just beyond our understanding. Roger Penrose, in his book *The Emperor's New Mind*, deals with the nature of consciousness and concludes that the answer may lie in quantum physics. He argues that the nature of thought transcends the limits of a mechanical computer, and so may require a form of 'hypercomputation' that exploits quantum uncertainty. His is a very sophisticated argument, but to my mind it is filled with vitalist desperation. He rests his hope on yet another poorly understood physical phenomenon to explain and contain life or, in this case, consciousness.

The ogre from which the vitalists have been running ever since the rise of rationalism in Galileo's day is the doctrine of materialism. This is odd, because vitalism itself is largely a materialistic notion, since despite its best intentions it tries to embody a spiritual idea in an essentially physical form. Materialism holds that there is no separate division into material and spiritual worlds. So there is no magical soul, no life beyond death, no need or room for a god. Everything is subject to the laws of physics and everything that exists belongs in the physical domain. This is perhaps a bleak outlook as far as the soul is concerned, but its success over the past three hundred years cannot be denied. From the time the great scientists of the Renaissance banished the Hand of God by showing how the planets move according to clear, unvarying rules, the universe has been gradually reduced to clockwork. When Mary Shelley first dreamed up the story of Frankenstein she was in Switzerland, sharing a house with her husband and his friend, Lord Byron. Byron was the father of Ada, Countess Lovelace, and it was she who so clearly saw and explained the implications of her friend Charles Babbage's inventions. Babbage used clockwork as the basis of his unfinished Difference Engine and, later, his even less complete Analytical Engine. These were the forerunners of the modern computer – a machine designed to 'think' – and it is perhaps this invention that has done more than anything to reduce the human spirit to a simple mechanical process.

I would like to assert that, although the materialist viewpoint is undoubtedly the truth, it is not the whole truth. I am a computer programmer by background, and as familiar as anyone with the means by which apparently abstract ideas can be reduced to simple mechanical steps. But I believe that the computer, if interpreted correctly, can be the saviour of the soul rather than its executioner. It can show us how to transcend mere mechanism, and reinstate a respect for the spiritual domain that materialism has so cruelly (if unintentionally) destroyed. The modern, fast, digital computer can give us a new understanding of the world of software, and this puts the world of hardware into a new perspective.

Save our souls

The beauty of materialist, mechanist, reductionist thinking is that it can explain so much about our world; its danger is that it can also explain things away. If you had just seen an impressive conjuring trick and I told you how the trick was done, I would be guilty of explaining it away. The beauty of the trick depends on it being inexplicable and magical, and reducing that magic to mere sleight of hand would trivialize it and spoil it for you. On the other hand, if you were enjoying a beautiful landscape painting and the artist began to explain to you how she had used light and shade, pigment and brush stroke, to fool the eye and suggest a feeling of depth or poignancy of mood, you would not find the painting any the less beautiful for this elucidation. Indeed, your increased understanding might considerably enhance your appreciation of it.

Science usually aims to achieve the latter effect, and scientific theories are advanced in the hope that they will enhance the beauty of the phenomena they seek to explain, rather than diminish it. Sadly, science also occasionally explains things away – perhaps inadvertently, sometimes deliberately. Often the difference comes down to a poor or malicious choice of words, especially the insertion of qualifiers such as 'just' or 'simply', as in 'consciousness is simply a product of nerve impulses'. To a large extent, philosophical speculation about life and consciousness has been a sad, gradual process of 'explaining away' the soul. It has been turned into first a substance, then a force, and now a mechanism. Yet recognizing that life and consciousness are products of material processes need not 'explain them away' in the slightest. In fact, I think it calls into question the meaning of 'material process' itself.

Only a fool or a coward would want life to retain its mystery as if it were a magic trick. Nothing is lost and much is gained by understanding things, and life should be no different from anything else in this respect. Yet we cannot allow these explanations to diminish us, as living beings. The poetry of life must not be replaced with matter-of-fact prose.

We have good reasons to be protective of our souls. Quite apart from our fear of mortality, we rely on our veneration of life to guide our everyday choices. Our division of the world into the categories 'living' and 'non-living' seems to be one of the most fundamental judgements we make and, whether it is fair or not, we treat each category in very different ways. Perhaps the most profound distinction we make between living and non-living is in our application of morals. The concepts of 'right' and 'wrong' are applicable only to living things. We never accuse an avalanche of being a murderer, and we never campaign for the rights of hurricanes. We are sure that killing other human beings is wrong, and some people are so clear on this point that they are even willing to kill the killer, in pursuit of some concept called 'justice'. We are less certain about whether it is wrong to kill a cow or an ant or a bacterium, but we generally assume that it is merely less wrong (or at least more defensible) as we move down some assumed hierarchy of soul-hood. On the other hand, we seem to believe that there is a sudden, qualitative change at the division between living and non-living, so we never trouble ourselves for a moment about the welfare of rivers or hills.

If life is reduced to mere clockwork, where does that leave our sense of morality? In fact, as life has indeed begun to be reduced to clockwork, and especially as we have gained mastery over that clockwork, so has our moral certainty declined. Today we face difficult moral judgements about abortion, euthanasia, genetic manipulation and mind control, which arise because our ability to understand has failed to keep pace with our ability to act.

But moral and ethical quandaries aside, I believe there is a sense in which our understanding of the universe has become a little inside-out. The primitive, dualistic distinction between mind and matter, spirit and substance, has been replaced by a more modern one which in effect claims that there is only substance. I would prefer to claim that there is only spirit. One of the points of view this book seeks to explore can be stated as follows:

> Life is not made of atoms, it is merely built out of them. What life is actually 'made of' is cycles of cause and effect, loops of causal flow. These phenomena are just as real as atoms – perhaps even more real. If anything, the entire universe is actually made from events, of which atoms are merely some of the consequences.

Some of these ideas will, I hope, emerge as you read the following chapters. In any case, I want to try to show that life is more than *just* clockwork, even though it is *nothing but* clockwork.

Applying the ghost in the machine

Life has always been an ineffable mystery, and many people under-standably prefer to keep it that way. Science has steadily been eroding this mystery, and sometimes it seems that the poetry of our souls is about to dissipate in a haze of prosaic logic. This book sets out to do two apparently incompatible things: to retain and reassert life's majesty while explaining how we can create it for ourselves. Part of my purpose is to show that life is more than the sum of its parts, yet at the same time I want to show that it is possible to understand these parts and assemble them afresh to create new life. This fact has more than mere curiosity value, and there are better reasons to play at being god than simply self-aggrandizement. Living, thinking, caring, motivated beings are very different from other classes of machine, and there are strong technological as well as social reasons why an understanding of life and mind is important.

Once upon a time, all machines were integrated with living things. Every plough was pulled by oxen and guided by a man; every lathe was turned by hand and controlled by the eye. The Industrial Revolution removed the need for muscle power, and to some extent the progress of automation has reduced our reliance on human supervision for the control of machines. Yet many jobs can't be done, or are done badly, without an intelligent living organism at the helm. A tractor can pull a larger plough than a team of oxen can, but unlike the oxen it cannot refuel itself or navigate rough terrain without an external brain to guide it. Since oxen are notoriously bad at driving tractors, this dull and rela-tively mindless duty fell to the far more adaptable but rather overquali-fied brains of human beings. Research into artificial life* is inspiring a new engineering discipline whose aim is to put the life back into technology. Using A-life as an approach to artificial intelligence, we are beginning to put souls into previously lifeless machines – not the souls of slaves, but of willing spirits who enjoy the tasks they are set. The first great age of technology can be thought of as the Inanimate Age, during which hammers and ploughs needed separate biological

---

* When I say 'artificial life' I am referring to synthetic living entities. This is rather different from the term 'A-life', which refers to a scientific discipline. A-life studies lifelike processes in the broadest sense (for example the mathematical simulation of evolution) but seldom con-cerns itself with attempts to create complete living creatures. Since this book is primarily about artificial creatures, I shall use the term A-life only when it is most appropriate.

controllers and power sources. After that came the Animate Age, in which steam replaced muscle and rudimentary automation took animals (if not people) out of the loop. The third great age of technology is about to start. This is the Biological Age, in which machine and *synthetic* organism merge. Taking our technological inspiration from biological systems promises to deliver a more adaptable, intelligent, robust and, above all, personable class of machines. Whether removing the need to employ profoundly intelligent human beings on mind-numbing and tedious jobs will alleviate the suffering in the world or exacerbate it is perhaps a moot point. Nevertheless, it will be good to have the choice, and undoubtedly there are many situations in which biologically inspired machines could improve or protect people's lives.

Seeing how the trick is done

After years of idle experimentation with the ideas to be described in this book, my first real opportunity to create artificial life came in 1992, under the cunning disguise of writing a computer game. This game was called *Creatures**, and it turned out to be more successful than I could ever have imagined. Roughly a million people across the world now look after the artificial organisms I created, and the global population of the creatures themselves even outnumbers that of many natural species.

One of the big surprises for me was the large following of devoted enthusiasts the product attracted. Many of them (quite rightly) regarded *Creatures* not as a computer game but as a scientific hobby, and some of their activities are astounding in their own right – so much so that some of the most experienced artificial life experts in the world today are probably non-academics who bought this game and took it and its creatures apart, piece by piece. I hope many of these enthusiasts will read this book. Perhaps you are one of them.

A game it may have been but, if you'll forgive the staggering lack of modesty this implies, *Creatures* was probably the closest thing there has

---

* Creature Labs and CyberLife are registered trademarks and the Creature Labs, CyberLife, *Creatures 2*, *Creatures 3* and *Creatures Adventures* logos are trademarks of CyberLife Technology Ltd in the United Kingdom and other countries. CyberLife, Albia and Norn are also trademarks of CyberLife Technology Ltd which may be registered in other countries.

been to a new form of life on this planet in four billion years. At the time of writing (and ignoring my present research work), these creatures probably still represent the state of the art in synthetic life forms.

Understandably, many people have since asked me to 'reveal the secret' of how to create artificial life, either out of scientific curiosity or even because they wanted to produce something like it for themselves. Trade secrets aside, I've generally been happy to explain how the trick is done, and have described many of the details in lectures, papers and articles for both scientific and lay audiences. But the facts alone are not really enough. I can and will tell you how it was done, and I'm sure you will follow what I am saying without any difficulty at all. If the facts were all you wanted, you could simply skip to Chapter 11 now and avoid all the philosophical stuff completely. But if you did, would you *really* understand it?

Maybe; maybe not. In my job as senior programmer and later as technical director of the company that made *Creatures*, I frequently tried to explain how the trick was done to some very bright programmers, many of whom are trained in A-life and far better qualified than I am. Yet I met with only limited success. This is not because the ideas are inherently difficult, or because the system I developed is especially complex (although it's not exactly simple!). No, for many people the reason is that to understand life, both natural and artificial, they have to change the way they look at the world.

I certainly had to think about a lot of interesting things in order to be able to write *Creatures*. But perhaps more important is the stuff I learned while I was writing it – the insights I gained from the experience. The problem with insights is that they are like skills: you can't use language to transfer them from person to person. I can tell you how to operate a video recorder, because that is knowledge. But even if I knew how to do it myself, I couldn't tell you how to be a better gymnast because that is a skill. Skills can be learned only through personal experience, and insights are much the same. Where new knowledge is simply added to one's existing mental store, insights bring understanding, and understanding changes one's whole being.

The act of writing *Creatures* – of trying to solve for myself many of the problems that Nature has had to solve through evolution – taught me many nameless things about life. It was by direct personal experience that I saw how Nature's trick is done, and it changed my understanding of the world profoundly. Unless you plan to shut yourself away for five years like I did, you may never discover these things for yourself. So I want to try to give you some of these insights vicariously instead. As you read, please remember that insights cannot be coded in prose. They are not things that you can simply take in; they have to be *felt*. What I'm trying to get across in this book is not so much a series of facts or opinions but a way of looking at things. You may have to read between the lines.

In a nutshell, the view I want to convey in this book is largely an anti-reductionist, anti-materialist and (to a degree) anti-mechanist one. The argument I want to put forward is that the natural world is composed of a hierarchy of 'persistent phenomena', in which matter, life, mind and society are simply different levels or aspects of the same thing. I also propose that this natural hierarchy can be mirrored by an equivalent one that exists inside a parallel universe called cyberspace. I then want to sketch an outline for a common descriptive language which can be used at all levels of the hierarchy. In this language we shall find the basic operators of which life and mind are constructed. To create artificial life we have to understand the nature of this hierarchy, implement simulations of these basic operators using a computer (or other device) and build upon that foundation the higher levels of persistent phenomena that we seek. A computer cannot be intelligent, but it can create a parallel universe in which natural forms of intelligence can be replicated.

Playing God

In summary, I am an aspiring, latter-day Baron Frankenstein. Like him, I believe that life can be created where there was none before. Like him, I think that it is possible to make thinking, caring, feeling beings and that, when these beings exist, it may be reasonable to ascribe to them a soul. Like him, this is what I have set out to do. Frankenstein's terrible and ultimately fatal mistake was to carry out the act of creation first, and to think about the consequences afterwards. Mary Shelley made him suffer for his impudence and his arrogance. Perhaps she was right, maybe it is arrogant to attempt to 'play god' in that way. To do so with the aim of debunking and debasing life would certainly be arrogant. But I think that there is a kind of under-standing that can be achieved only by building things for oneself, and that this kind of understanding generates a respect for the subject that may be lacking in one who merely 'sees how the trick is done'. So far, my faltering attempts to create life have only increased my admiration for it. If we succeed in grasping the mechanism of life, as we have grasped electricity and chemistry, then, rather than familiarity breed-ing contempt, I suspect we shall find a new, harmonious understand-ing of everything, and life will be elevated once more to its proper place.

* *

CHAPTER ONE

FAILING THE TEST

The second woe has passed; behold, the third woe is soon to come.

Revelation 11: 14

Today is the first day of a new millennium, and the date on my laptop reads a comforting 01/01/00.

Outside, the misty prickling rain that for England marked the passing of the twentieth century has given way to a calm, fresh-start sort of a morning. A watery sun presides over a silent, clean-washed landscape, like a painter who has prepared her canvas, laid out her tools neatly and methodically, and is taking a last deep breath and a moment's reverent silence before setting to work on a new creation.

Yesterday also marked a minor defining moment in my own life, for reasons I won't trouble you with, but it drew a neat symbolic line under an otherwise rather turbulent year. Suffice it to say that I, like many people across the globe (at least, those without hangovers), am sitting here in the wintry sunshine, calm and centred, waiting for the new beginning.

The year 2001 bug

Yet it wasn't supposed to be like this. Had the soothsayers had their way, midnight would have ushered in a more eerie calm, punctuated only by the wail of sirens, as the world's computers ground to a halt and civilization collapsed in an embarrassed heap. Happily, the worst of the much-touted millennium bug, which was to perplex date-calculating software by appearing to make $99 + 1 = 0$, has failed to cause the damage many people feared it would, and we can sleep soundly in our beds a little longer.

But the year 2000 bug, according to the soothsayers, was really only the beginning of the end. If we are to believe Arthur C. Clarke, we have very little time left before computers start locking us out of our spaceships and intelligent machines take over the world. What we should be afraid of on this bright new year's morning is apparently not Y2K but what we might call the year 2001 bug – for that is the year in which, according to Clarke's and Kubrik's haunting *2001 – A Space Odyssey*, the electronic brain of the computer known as HAL 9000 will take cold logic to its worst extreme and commit murder.

Of course, we should always believe the soothsayers. If you grew up, as I did, with tales of Dan Dare and Flash Gordon, you will know that up above our heads right now is Moonbase Alpha, and the streets outside are full of people wearing natty little silver-lamé spacesuits with flared trousers and matching holsters for their ray guns. The universe is also apparently overflowing with alien races whose only object in life is to destroy our pathetic little planet. And if they don't get us, the mechanoids will. Out there in the crystal city, under swarms of personal helicopters and illuminated by the searing light of nuclear fusion, are anonymous grey office blocks containing giant intelligent computers, which can easily out-think us and even now are conspiring to bring about our destruction by coordinating their vast armies of androids.

If you are younger than me, your memories will be of *Star Wars* and *Terminator 2*. It doesn't matter: the fashions will be different but the plot remains the same. Machines with lives and minds of their own are clearly something to fear.

Perhaps happily for us, however, New Year's Eve 1999 also marked another significant watershed. Fifty years earlier, a brilliant young man made a plausible and worthy prediction, one on which many of the soothsayers' tales of doom depend. But time ran out for him yesterday, and he was proved wrong. That man was Alan Turing, one of Britain's greatest unsung heroes. Turing did the lion's share of inventing the digital computer: not only the fundamental theoretical work but also a detailed practical design for a machine, and even the world's first instruction manual for computer programmers. This was a huge scientific and engineering achievement, and yet, because much of Turing's working life had been cloaked in wartime secrecy, and not least because he was hounded to an early death for being gay, many people have never heard of him.

Turing's interest in logic and the construction of computing machines stemmed from a childhood fascination with the human mind and soul. He thought a great deal about the process of thought itself, and his work on mathematical logic and codebreaking machinery led him to conceive of a machine that could think. To him, the digital computer was not some kind of giant calculator or glorified typewriter, but a logical machine – and therefore a reasoning, thinking machine. Turing was an early member of the cybernetics movement (of which more later). He was also one of the founders of artificial intelligence (AI). If the term 'A-life' had been coined back then, Turing would have been a pioneer of that too, because intelligence was only one of the attributes of life that he sought and studied.

To Alan Turing, the workings of the human mind and the mysteries of biological growth and development could be visualized as step-by-step processes carried out by a complex machine – a living organism. Knowing what he knew from his wartime experiences with automatic codebreaking equipment, and armed with a unique mathematical background in step-by-step mechanisms, Turing, along with other pioneers, set out to replicate some of the processes of life inside a digital computing machine. By 1950, he was willing and probably uniquely qualified to make the following prediction about the chances of success: 'I believe that at the end of the century ... one will be able to speak of machines thinking without expecting to be contradicted.'

Well, the end of the century has come and gone, and anyone who uses computers on a regular basis will know that they are incredibly stupid machines that cannot even count the date properly. Research into AI has given us ingenious machinery that can sort nuts from bolts, diagnose diseases from their symptoms or recognize whether somebody is smiling. It has also given us the Windows 'paperclip', for which it will never be forgiven. But has it given us machines that really think in the way that living things do? I fear not.

Whatever happened to HAL?

The one thing Alan Turing gave the world that the world actually still remembers is his famous Turing test. He devised an experiment to judge whether a hypothetical computing device could fairly claim to be intelligent. In the test, an operator conducts two conversations by typing questions into a terminal. One conversation is with a human being and one is with the machine, but the operator is not told which is which. If the operator cannot tell which of the two is the human, the computer is deemed to have passed the test. Such a test has actually been carried out a number of times now and, perhaps surprisingly, some AI programs have been sufficiently convincing that the operator has been fooled, at least for a while.

Simple stored sentences, regurgitated automatically in response to certain key words in the question, can quite easily fool people for a short time. But this is like assuming that a book of multiplication tables can actually multiply. Ask the tables a question beyond their limits, or conduct a conversation with a computer program for long enough, and you can see that regurgitating stored knowledge on cue is not the same thing as intelligence.

Here is another little test of intelligence that I find more appealing. Turing and the early AI pioneers frequently cited the ability to play chess as a test case for intelligence. So, imagine a high-powered AI chess-playing computer, like IBM's much famed Deep Blue. Also imagine a rabbit. Now try to visualize what happens if the rabbit is asked to play chess against the computer. It turns out that rabbits are really not very good at this – the queen's opening gambit gets them every time, for example. On this reckoning, Deep Blue is very much smarter than a rabbit. But now imagine dropping them both into a pond. In my view, the one that is really the most intelligent will be the first to figure out how to avoid drowning!

Intelligence involves a great deal more than the ability to follow rules (which is what a chess-playing program does). It is also the ability to make up the rules for oneself, when they are needed, or to learn new rules through trial and error. It is true that chess computers are handicapped by their lack of any means of propulsion, so that in the above scenario drowning is, for them, the only option. Nevertheless, even if Deep Blue had been given flippers it could not save itself unless its designers had explicitly programmed it to swim and told it when to do so. The intelligence would thus belong in the minds of the programmers, and only the end result of that intelligence, encoded as a set of explicit rules, would reside within the computer. Rabbits, on the other hand, will recognize the warning signs of imminent doom, try an assortment of movements, and quickly learn to repeat and perfect any actions that seem to help. Life finds a way to survive. Computers simply drown, and they neither know nor care that they are doing it.

So modern computers, even when programmed by AI experts, are really not very bright. 'Smart' might be a better word, but I think to call them intelligent is just debasing the term. People don't yet routinely talk of computers thinking, except in a metaphorical sense. In a way, fifty years of AI research finally failed its own Turing test last night, on 31 December 1999, when Alan Turing's prediction ran out of time. Why was this? Well, it was certainly not Turing's own fault. He was a brilliant man, with thoughts far ahead of his time. Even many computer scientists don't know that he also experimented with other forms of computation that might, had he lived, have had a far greater influence than the digital computer. But Turing and his wonderful machine started people travelling down a path that led the wrong way.

Paradoxically, part of the reason that AI has failed so far is its very success. The problem with all fields of research is that people are impatient. If a particular line of enquiry seems to be making progress, we continue down that line, but if it seems to be getting nowhere, we abandon it. This is a problem, because the route to the future is often tortuous. Things seem to be moving towards the goal but then unexpectedly snake off in the wrong direction. Initially unproductive approaches can often turn out to be the only ones that lead to the desired destination. Turing, for example, had three brilliant ideas about computation. These three might be characterized as 'organized machines', 'unorganized machines' and 'self-organizing machines'. My feeling is that the last two hold a great deal more promise than the first, but that first idea was so stupendously successful that it eclipsed the others more or less completely for nearly half a century. Turing's unorganized machines are what we nowadays call neural networks, while his self-organizing machines explored one of the processes that may help to explain how a simple, undifferentiated egg cell grows into a complex adult organism. The organized machine, which set the tone for the study of lifelike and mind-like processes for years to come, was the digital computer.

I really have nothing against computers. There are half a dozen personal computers in my house and countless microprocessors working behind the scenes and I adore every one of them. I've spent twenty-five years programming them and can remember the microprocessor when it was but a mere lad in short (4-bit) trousers. The very idea of computers – the look of them, the culture associated with them and above all their amazing capacity to create whole other universes out of simple arithmetic – all these are intoxicating. The digital computer is the most masterly invention of the twentieth century, if not the second millennium. But it is still an organized machine. It was a tractable idea that showed great success in the early stages (say the first forty years), but as far as making living, thinking beings is concerned, it is a bit of a dead end.

In essence, the problem is that the digital computer was modelled on the outward appearance of mental processes, rather than the structures that give rise to them. Even though we know our brains consist of vast numbers of neurones operating in parallel, we each appear to have only one mind. This mind seems to operate in a stepwise way, thinking about or carrying out sequences of actions one at a time. We also get a sense that our conscious thoughts are at the top of a chain of command – we take the big decisions consciously, but then delegate the task of carrying them out to some lower, subconscious parts of our brains. The mind therefore gives us the impression that it is top-down (employs a chain of command), serial (only one mind per brain, operating one step at a time) and procedural (works in terms of logical procedures to be followed, as in a recipe). The digital computer is similarly a serial machine because it only carries out one operation at a time. It is procedural because the basic units of a program are actions to be carried out (such as 'add these two numbers and store the result here'). It is also top-down, since computer programmers tend to design their programs as control hierarchies – a central program carries out commands by issuing orders to subroutines, which in turn invoke subordinate routines to handle the finer details.

The computer was designed as a model of how the mind seems to work, and the operation of a computer program was assumed to be very similar to thinking. Yet there are flaws in this logic. For one thing, there is a potential pitfall with the top-down approach. In any chain of command the buck stops with the person at the top, and with a computer program it is all too easy for the buck to stop, not with the top level of the program, but with the programmer. In other words, what seems like intelligent behaviour initiated by the machine (for example the ability to play chess) is often just the stored intelligence of its designer, regurgitated on cue. Also, it doesn't follow that copying the outward appearance of something is the same as recreating the thing itself. Statues are not people: they just look like them. Sooner or later, the mask is bound to slip and the deception will be exposed. Finally, it is really only philosophers and mathematical logicians who would believe that thinking amounts to the formal manipulation of symbols according to set rules. Most of the time the rest of us don't think in neat syllogisms or conduct formal arguments in our heads. More often than not the answers just occur to us in some mysterious way, and we use logic only in retrospect as a means of justifying our conclusions to others or to ourselves.

So the digital computer was in many ways the wrong tool, applied to the wrong job. Ironically, though, this most organized of machines is such a powerful concept that it can actually get around its own limitations, but only if one thinks about it in the right way. This book is very much about how to turn the prim, tightly organized digital computer into a disorganized, self-organizing machine. We shall use the serial, procedural, top-down computer as a tool to create new machines that are parallel, relational and bottom-up. In this direction lies the goal that Turing sought, and I hope he would have approved.

The art of stating the painfully obvious

I suspect that the early pioneers of lifelike artificial systems were really only reflecting the spirit of their time. They began their work immediately after the Second World War, in an environment dominated by huge, top-down military organizations and a political environment based squarely on command and control. There was still a rather Victorian attitude to technology, in which machines were seen as a way to dominate and conquer nature. The science of the period was also very elemental, trying to pare the world down to its bare essentials and hoping to explain everything in terms of almost nothing. What is more, these people had grown up in the 1930s, a time of utopian dreams when a sterile and impersonal brave new world actually seemed like a good idea.

Today we are beginning to see things rather differently. Top-down is giving way to bottom-up as corporations downsize and outsource, totalitarian states crumble, and the Internet begins to make democracy and individual control a reality. Our fear that we are damaging the environment has changed the way we view our relationship with the natural world, too. We are starting to learn how to work alongside nature rather than against her. Even our science is changing, as chaos theory and complexity theory help us to understand whole systems in ways that we never could when we looked only at their parts. So top-down is being replaced by bottom-up, procedures are now seen as less important than relationships, and our ability to cope with things that are complex and parallel (paradoxically, thanks to the invention of the computer) reduces our desire to serialize and simplify them. In short, I believe we are undergoing what Thomas Kuhn called a paradigm shift.

We are starting to look at the world in a different way, and the consequences of such a change of viewpoint can be profound. Paradigm shifts take place when people start to question their previously unspoken basic assumptions. For example, Isaac Newton helped to precipitate a paradigm shift when he pointed out a few basic facts about how objects behave when they are moving. The resulting theory of mechanics changed the world immeasurably. These things appear blindingly obvious to us now, and it seems inexplicable that no one thought of them before. After all, Newton's world-shattering laws of motion can be paraphrased simply enough:

One: if you push something it will keep moving until something stops it.
Two: the harder you push something, the faster it will accelerate.
Three: it hurts when you kick things because they kick you back just
   as hard.

Why it took centuries for anyone to work this out is a mystery. Yet once ideas like this had been stated explicitly, they revolutionized the whole of human thought and changed all our lives. But that's how it is with paradigms – you don't realize when you're stuck in one. For example, if your whole world revolved around the idea of divine intervention and the notion that the way things behave is up to God to decide, then having someone like Galileo stand up and insist that things universally fall at the same rate, regardless of what they are made of or what God intended them for, was bound to be a bit hard to swallow.

In many ways, philosophy is the art of stating the obvious, but until someone actually stands up and says so, most of us continue to leave our assumptions unquestioned. The unquestioned assumptions that underlie the past fifty years of AI research, not to mention several centuries of thinking about the nature of life and mind, are many and varied. Yet things are starting to change.

I have listed below a few of the assumptions that I think some of us are at last beginning to call into question. I believe that our understanding of life, of what we are and of how to build artefacts that share these properties, will only progress when we abandon the old axioms and learn to look at the world in a new way. The rest of this book is a series of linked essays that I hope will help to shed a glimmer of light in the right direction. As I stand here on the first morning of a new millennium, the early light of a new way of thinking about things is already starting to break, not just in AI or neuroscience, but also in politics, economics, engineering and just about everywhere. The twilight stretches back into the past century, but in Turing's day the way forward still lay in shadow. We stand now on the verge of a new century and a new paradigm. But, like all paradigm shifts, first we have to learn to let go of some of our most cherished and unquestioned notions.

Conventional wisdom	*My contention*
Minds can be explained in terms of physics.	Physics does not even understand matter properly, and is ill-equipped to understand mind.
Computers can be intelligent.	Computers can create spaces in which intelligent things can be built.
Control is synonymous with domination.	Control is as much an effect as a cause, and the idea that control is something you exert is a real handicap to progress.
Intelligent systems must be designed from the top down.	Intelligent systems must be designed to emerge from the bottom up.
Intellect and intelligence are roughly the same thing. The ability to reason can be separated out and implemented in isolation from other modes of thought.	Intelligence is first and foremost about common sense. Reasoning (which is only one of many aspects of intelligence) must be built upon a foundation of common sense.
Intelligence is independent of life.	A system will not be intelligent unless it is also alive.
Intelligence is a unified process and can be implemented directly as an algorithm – a sequence of logical steps.	Intelligence is a property of populations. Although we seem to have a single stream of consciousness, it can be reproduced only through a parallel process.
The best way to design a machine that thinks is to examine the process of thought.	The best way to design a machine that thinks is to examine the structure of biological systems and the behaviour of mechanisms that lie much deeper than conscious thought.

* *

CHAPTER TWO

LIES, DAMNED LIES AND LINGUISTICS

'When I use a word,' Humpty Dumpty said, in a rather scornful tone, 'it means just what I choose it to mean – neither more nor less.'

Lewis Carroll, *Through the Looking Glass*

Most books that refer to the brain and mind include at least one picture of an optical illusion to illustrate some point or other, and I wouldn't dare to break with this tradition. But my purpose in including Figure 1 is somewhat different – perhaps even the opposite of the usual one. The reason that this optical illusion is startling (and hence remarked on in psychology books) is that we cannot decide whether it is a picture of a vase or of two faces. There is nothing to choose between them; each is equally probable and familiar. This is so troubling because it is in marked contrast to our everyday experience, where there is a distinct difference between one interpretation and the other – between figure and ground. The event shown in Figure 2 never really happens. We call Figure 1 an optical illusion because of this paradox, while we take our everyday experience to be no illusion at all. The purpose of this chapter and the next, however, is to claim that all along we have been looking at the natural world the wrong way round. What we perceive as the 'correct' way of looking at things is actually mistaking the ground for the figure. To understand life, mind, consciousness and soul, I believe we have to learn to turn our intuitive interpretation of the world inside out. In short, we need to stop looking at the actors and instead start focusing on the play.

What's in a word?

Left-handed scissors must have been invented by a right-hander. How

Figure 1. Is it a bird? Is it a plane?

anyone thought that it was a good idea simply to attach left-handed handles to right-handed blades is a mystery to me. Perhaps it never occurred to anyone that scissor blades are handed in the first place, but they are. Because the top blade of a pair of scissors lies to the left of the bottom blade, a right-handed person's thumb naturally forces the blades tighter together as they cut. But hold the same scissors in your other hand and, no matter how beautifully the handles have been reshaped to fit the left thumb, the result is very different. Closing the blades now causes them to spring apart instead of gripping together, and whatever it is they are supposed to be cutting twists between them but remains uncut.

Figure 2. This never happens.

Have you ever noticed what a grudge the world has against left-handers? If you are right-handed you might not have detected it but, as a 'cack-handed lefty' myself, I find it remarkable that words to do with the left generally have negative connotations, while those to do with the right are positive. The Latin word for 'left' is *sinister*, while *dexter* means 'right'. Sinister is nowadays a bad thing to be, while to call someone dextrous is a compliment. We insult people by calling them 'gauche' (French for 'left') or compliment them on being 'adroit' ('right'). Even within English itself, the word 'right' is synonymous with 'correct'.

Yet there is an equally odd and perhaps far more influential bias in the way language distinguishes between another set of opposites, this time between the tangible and the intangible – between matter and (for want of a better word) form. A company's 'tangible assets' are apparently good things to have, while their intangible ones are somehow inferior. 'Substantial' is a positive thing to be, while 'insubstantial' is not. Reliable people are 'solid', and my thesaurus tells me that 'concrete' is roughly synonymous with 'real'. Things that really matter are 'material' (as in 'material facts'), while those that don't matter are regarded as 'immaterial'. Even the word 'matter' in the previous sentence is pejorative, for there is a definite underlying assumption that matter is real and good, while the relationships between these material things are somehow not, and this bias is deeply embedded in our linguistic heritage.

Language provides an important part of our toolkit for conscious thought. For many of us it is very difficult to think consciously without speaking words in our head. Without the right tools, it is very difficult to do an effective job – try cutting paper with left-handed scissors and you'll know what I mean. We can sometimes find it very difficult, if not impossible, to think certain thoughts, simply because words for certain things either don't exist or carry inappropriate baggage with them. The reason why we esteem the material world more than we do the intangible one is fairly obvious – it is the world that our senses tell us is really 'out there'. Our eyes see physical things, or so we like to believe (but remember that we 'see' only the effects of visible radiation emitted or reflected from what we take to be solid objects; we don't really see the objects themselves). Our touch sensors tell us when we have bumped up against a solid structure, and our ears reliably report the interaction between distant vibrating objects and their surrounding air. On the other hand, we do not have any direct sensory confirmation of intangible things. We don't have poverty sensors, we cannot touch a society, and our only evidence for the existence of other people's minds is the visible or audible motion of their physical bodies. Consequently, we come to believe that the things we can directly sense are real, while the things we cannot sense are more like figments of our imagination or convenient labels, rather than anything absolute, independent and genuine.

And yet despite all this, the things we really care about are largely intangible. 'Life' is an intangible concept, as is 'mind'. We care about suffering in a way that we never do about mass. This has led to some strange and almost perverse logical errors in the past. Even today, despite some major changes in our perception of the intangible (for example our familiarity with the concept of software), our outlook limits our ability to deal with and understand our world and ourselves. As I mentioned in the Introduction, the modern, scientific, materialist viewpoint sometimes forgets its noble intention to explain things, and instead 'explains them away' through sloppy use of language. To say that the mind is simply the product of nerve impulses is literally correct. Nevertheless, because of the pejorative use of 'simply' and the unconscious value judgement placed on material things such as neurones, when compared with immaterial things such as minds, this statement demeans and almost explains away the mind altogether. If this weren't bad enough, the alternative, supposedly opposing school of thought gets even more trapped by our linguistic materialism. The vitalist school claims that mind and matter are distinct and different, and that each is equally real and valid. We shall look later at the question of whether mind and matter are distinct and different, but I would certainly applaud the spirit(!) of the dualist viewpoint. At least it attempts to validate the reality of life and mind in a way that materialism fails to. Sadly, this dualistic notion is just as badly served by the concepts permitted by our language as materialism is. Far from treating body and spirit as conceptually distinct, vitalism attempts to reify the spiritual world and treats it is as some special form of substance. It is as if spirit can be made real only by 'promoting' it into the material world, and treating life and mind as if they were some special kind of fluid.

I am finding it very difficult to choose words that don't carry materialistic connotations, and even more difficult to find suitable language to express what I mean by 'the intangible', and I think this probably proves my point. Our ability to reason is conditioned by our language, which in turn is conditioned by the evidence of our senses. It is hard to break free from our innate respect for 'stuff', yet until we do this we shall never understand life, because life is an intangible thing.

There's no such thing as a thing anyway

We need to reconcile mind and matter, and for that we must carry out some thought experiments and look at some examples to convince us that mind is neither inferior to matter, nor a special example of it. But let us limber up a little first, by playing with our cosy but not always helpful or accurate notion that the world is made of 'things'.

The visual signals that enter our brains when we look at a scene arrive independently, in parallel, like the individual dots of ink in a magnified newspaper photograph. Yet by the time we become conscious of it, the visible world has been carved up by the brain into discrete objects. I see a table and a computer in front of me now, not a grey blob in the middle of a brown one or an array of unconnected points of light. The retina of my eye and the visual cortex of my brain have enhanced the edges between blocks of colour and detected lines of various orientations. Individual features have been assembled into discrete objects, which my memory has identified and labelled as 'computer' and 'desk'. No matter how certain I am at an intellectual level that my computer is a cloud of tiny atoms, separated by relatively vast regions of empty space, my brain tells me that what I am looking at is a single thing. This is just as well, or I would never be able to make sense of my world and interact with it. To all intents and purposes, my computer can be treated as a single entity – I can touch and manipulate its component atoms as a unit, and the fact that I think of this grey blob as a computer enables me to interact with it and produce this book. Nevertheless, we can easily forget that what we perceive is not necessarily that which is really there.

Putting boundaries around things is an important part of what brains do. We isolate and classify our world at many levels, and this is an essential part of intelligence because it enables us to organize our responses by forming categories of things to eat, things to run away from, and so on. But we sometimes fail to differentiate between finding where to draw the line and choosing where to draw it. We have all seen examples of fallacious 'black-and-white reasoning', where something that in reality forms a continuum is treated by some people as if it were split into two distinct camps, and warrants only two possible courses of action – good versus bad, us versus them, alive versus dead. At this more abstract level the error is usually easy to spot, although it is not always as easy to counteract. Yet even at a much lower level our brains' innate ability to divide and group the world into discrete objects without a moment's conscious thought can sometimes lead us astray, especially if we form the conclusion that things are things only when they are directly perceptible and localized in space.

Two new heads and three new handles

Let's look at a few everyday things and see how counter-intuitively insubstantial they really are. Figure 3, for example, shows a picture of a cloud. This is the particular kind known as orographic cloud, which sometimes gathers around the peaks of tall mountains during strong winds. But hang on – if there is a strong wind up there, why doesn't the cloud blow away? Clouds are made from nebulous veils of tiny water droplets that we might expect to be whisked away at the slightest breeze, and yet orographic cloud remains stationary just upwind of the mountain peak. The answer, of course, is that the cloud is not a *thing* in the way that my computer (apparently) is. What happens is that moist air is forced upwards when it meets the mountain's windward slope, and as it rises it cools. At some point it will have cooled enough for the water vapour to condense out as droplets, which we see as the leading edge of the cloud. These droplets career through the sky at the same speed as the surrounding air and then find themselves tumbling down the other side of the mountain. As they fall, they warm up and re-evaporate, and this event marks the leeward edge of the cloud. Because the points at which the water condenses and re-evaporates remain stationary, the cloud looks like a static, discrete thing. In reality, water molecules are moving from place to place, and momentarily becoming a cloud.

Figure 3. Orographic clouds.

This doesn't apply just to orographic clouds. The puffy, white, cumulus clouds that you see on a summer's day are constantly changing in a similar way, except that here the air is being forced upwards by convection, rather than wind. The vapour condenses at a certain height, moves up through the cloud and then re-evaporates as it begins to fall down the sides of the mushroom of convecting air. This is why clouds are such a paradox: we know that they contain many tons of water, and yet they float lazily over our heads as if they weigh nothing at all. In a very real sense, they *do* weigh nothing at all, since a cloud is just a name we give to a region of space, through which moist air passes and momentarily renders its water content visible.

Clouds are not unique in this respect. To take a frivolous example, if you dug a hole in the ground, and then repeatedly removed earth from one side and added it to the other, the hole would move along. Is it still the same hole? Is a hole a thing anyway, moving or not? Recall the story about the old craftsman who was very proud of his tools. 'I've had this hammer forty years, man and boy,' he'd say. 'Of course, in its time it has had two new heads and three new handles ...' Clouds and holes and hammers may seem like special cases, but in fact they are more the norm than the exception. Figure 4 shows an example from the living world. This is a jellyfish. Actually, let me rephrase that: this is a colony of jellyfish. No, wait ... Many simple animals are barely more than loose coalitions of mildly specialized cells, and it is hard to decide

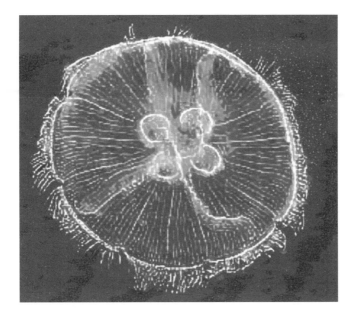

Figure 4. A loose confederation of cells.

which object should be referred to as an individual and which as a colony of individuals. Jellyfish, for example, live double lives – the familiar floating medusae reproduce sexually and give rise to a number of larvae, which then turn into polyps. These attach to rocks and live out an independent existence, reproducing asexually to create whole colonies of further polyps. Finally, each of the polyps gives rise to several new medusae. It is difficult to know when to talk about parent and child, individual and group, in such cases, and the whole phenomenon seems more like a nebulous cloud of constantly recycling biological tissue.

Finally, consider yourself. I want you to imagine a scene from your childhood. Pick something evocative. I'm remembering the pleasure of walking down the hill from my primary school one crisp, autumn day when I was about ten years old, kicking my feet through piles of flame-red sycamore leaves. This was perhaps the first moment in my life when I realized I was fascinated by complex dynamics – by the study of the intangible. I remember musing on the fact that autumn leaves always tend to congregate in the same places, often in great swirling whirlpools in the middle of the road. First of all, how did such light things manage to remain in such open spots when the wind was quite strong, and, secondly, was it a bunch of leaves whirling around in a vortex, or was it a vortex, whirling some autumn leaves?

Anyhow, by now I hope you have thought of an experience from your childhood. Something you remember clearly, something you can see, feel, maybe even smell, as if you were really there. After all, you really were there at the time, weren't you? How else would you remember it? But here is the bombshell: you *weren't* there. Not a single atom that is in your body today was there when that event took place. Every bit of you has been replaced many times over (which is why you eat, of course). You are not even the same shape as you were then. The point is that you are like a cloud: something that persists over long periods, while simultaneously being in flux. Matter flows from place to place and momentarily comes together to be you. Whatever you are, therefore, you are not the stuff of which you are made. If that doesn't make the hair stand up on the back of your neck, read it again until it does, because it is important.

The whole sort of general mishmash

To understand life and mind we have to learn to let go of our natural tendency to divide the world into discrete chunks. Living organisms are systems in flux, their constituent stuff changing from moment to moment; minds are not really 'things' in the conventional sense at all. But then, nor are clouds. All of these 'things' are shifting, blurred, interacting eddies in a single stream, and we must be careful when attempting to draw boundaries around them. In the next chapter, I am going to try to convince you that even matter itself doesn't really exist, and you will probably start to think that I must be a Buddhist. I am not, although I gather there are some similarities between my outlook and Buddhism. Neither am I a middle-aged hippie, but here's something else that might make you think I am.

If you live in the country, where the air is clear, you may sometimes have looked up into the starlit night sky and had the shock of developing a momentary sense of proportion. Nobody can really grasp what it means to look out across an infinite void, with millions or billions of kilometres stretching between you and the nearest object, but sometimes the Earth's atmosphere seems so thin that you can become quite giddy with the sense of emptiness out there. If you lean back far enough, or if you lie on your back, staring into the starry depths, then you can quite literally feel as if you are about to fall off the planet and hurtle out into the inky nothingness. It doesn't take a huge leap of the imagination to flip from knowing that 'down' is towards the ground to believing that it lies the other way, and you are dangling above a bottomless precipice. Stretching the imagination to give you such a momentary sense of proportion is a very good thing to do. Most of the time we are right to avoid such a sense – after all, nobody could carry on their everyday lives if they perceived, without attenuation, all of the suffering that is happening in the world. Nevertheless, a wise person will exercise their imagination whenever they get the chance, and developing the ability to visualize enormous distances is a valuable way to while away the odd moment.

Incidentally, you can do this in time as well as in space. Sometimes you can find rock outcrops that have been exposed along their bedding plane – the original land surface or seabed at some instant in the distant past. It can be very illuminating to sit on such rocks and ruminate on their great age, then to look down and see the impression of raindrops and the tracks of once-living creatures, fossilized in a momentary snapshot. There beneath you lies a visible record of a short shower of rain, one sunny Tuesday afternoon, about four o'clock, two hundred and forty *million* years ago!

I'm not going to insist that you lie upside down outside in the dark and try to believe you are falling off the Earth. But it can be good to try stretching your imagination to see if you can rid yourself of the brain's built-in tendency to carve up and categorize the world. It helps to see the world as it really is, not as our senses, our brains and our language fool us into thinking it is. Stare at the scene in front of you, perhaps out of a window. If it won't raise too many eyebrows, try looking at it upside down, so that your brain has more trouble recognizing the objects in front of you. Now attempt to stop believing in discrete objects altogether. Instead of seeing the separation between things, see the continuity between them. Try to grasp a sense that what you see forms a continuous surface, rather than a collection of separate things. If you can visualize it as a continuous surface, add the air and the rocks under the ground and try to imagine everything as a varying but continuous *volume*. Perhaps you will find it easiest if you explode everything into myriad tiny atoms in your mind – they occupy every space, but they can swirl around one another like the tiles in a sliding-block puzzle. Then try to add the fourth dimension, and feel how that volume flows and changes with time – how the regions of the scene that we think of as plants grow and fade; how they add to the air and take from it; how the rocky areas bend and buckle. It is a difficult state of mind to achieve, but I think it may help you to cope with the other ideas described in this book. The writer Douglas Adams, who has a tremendous feeling for such concepts, talks about 'the interconnectedness of all things', and this is the sensation I want you to seek. More accurately, I want you to get a feeling for the interconnectedness of all *events* – the ebb and flow of cause and effect. The episode from my childhood that I mentioned above was one of those rare and precious moments when I saw the world as a whole, not as distinct parts. I can only ever catch glimpses of this wholeness, but it is in these moments that I really grasp the nature of life, of what I am.

* *

CHAPTER THREE

A GUIDE TO THE INTANGIBLE

Everything is what it is because it got that way.

D'Arcy Thomson, *On Growth and Form*

Perhaps you are starting to see what this book is searching for. To say that you are not the stuff of which you are made is both true and false at the same time. To see yourself as a persistent *phenomenon*, when the substrate from which you are made is in constant flux, is to begin to understand life, and more than just life. Life is not a magical, fluid substance, but neither is it simply a convenient label to attach to certain combinations of material substances. In fact, material substances themselves are not even as substantial as we have been led to believe. Another thought experiment will, I hope, make this idea clearer.

Making a splash in Ireland

At the moment, I'm sitting outside a small, white cottage on the west coast of Ireland. Over my left shoulder I can see the purple, marbled mountains of Connemara tumbling into the distance. To my right, the grey Atlantic Ocean stretches away to infinity while, in front of me, beyond the rhododendrons and emerald grass and nestled among some low, rounded hills, lies a small, peat-black body of water known locally as Lough Maumeen. It just so happens that we need some water for our thought experiment. Just a bowlful would do, but why should I be content with something so prosaic when this mystic Celtic pool lies so conveniently before my eyes – and now yours?

Conjure up in your mind an ancient druid priest to enchant the waters. With a wave of his gnarled staff he can deform the water's

Figure 5. Defying gravity.

limpid surface at will. Imagine him raising up great pyramids and columns and glassy domes of frozen water, like Kubla Khan creating 'that sunny dome; those caves of ice' in deepest Xanadu (Figure 5). Then, as these glittering edifices stand quivering before us, he raises his arms slowly once more, pauses and then dashes them dramatically to his side. The spell is broken. Gravity is restored, and the crystalline citadel is smashed into a trillion tiny pieces. After a horrendous noise, as of clashing cymbals, comes silence. Where there were once examples of every kind of geometric and naturalistic form, there remain only echoing ripples and a few tiny whirlpools.

The point to grasp from this Irish daydream is that, even if you are able to fashion a vast variety of shapes in the surface of a body of water, within seconds all those shapes will disappear – all, that is, except for two. On the surface of water, sinusoidal ripples and swirling vortices are more or less stable forms, while domes and columns and pyramids are not. Herein lies an important point, and a central theme of my story. We have uncovered the most important law of nature, and it is this:

Things that persist, persist. Things that don't, don't.

I fully appreciate that this statement is a tautology, and also that it is blindingly obvious. But remember that philosophy is the art of stating the obvious, and just because it seems self-evident and even tautological now that I've said it, you shouldn't assume that it is not profound. In fact, this one statement explains all that you see around you.

What our lakeside thought experiment has shown us is that ripples and whirlpools persist for extended periods, while all other shapes disappear very rapidly (often becoming ripples or vortices themselves in the process). Given a random selection of shapes to begin with, the proportion of ripples in relation to other forms will increase over time, simply because the highly transient shapes disappear more quickly. Moreover, since many transient shapes turn into ripples as they collapse, ripples are coming into being throughout the event and, once made, they tend to persist.

Not only are ripples persistent phenomena but, just like clouds and people, they are examples of systems in flux. When we say a wave moves across water, we really mean that the water itself just moves up and down as a wave-like disturbance propagates across the surface. The thing we call the wave is not really made of anything, or at least not made of the same thing from one moment to the next, just as orographic cloud consists of water but not always the same water. A ripple is a pattern that persists by means of propagation – it copies itself forward in space and time. This ability to copy oneself is a clever trick, and we'll see other examples of this later. Anything that has a clever trick up its sleeve to enable it to persist will last longer than things that don't. So everything you see on the surface of water is likely to be something that possesses a clever trick for persisting. We can ask ourselves *why* a ripple persists (because it is inherently stable) and *how* a ripple persists (by copying itself forward in space).

Persistence is a virtue

If this kind of thing applied only to water, then it wouldn't be very important at all. But the truth is that it applies universally. *Everything* you see around you is an example of a phenomenon that persists, at least for a reasonably extended period, and the reason the universe looks the way it does is that some things persist and others don't. Rudyard Kipling wrote a poem about 'six honest serving men' named 'What and Why and When, and How and Where and Who'. Science can be seen as the process of asking these very questions about persistent phenomena. Naturalists may study *where* a phenomenon persists (a species of animal, say). Palaeontologists look at *when* things persisted, and thus how one form of persistence gave rise to others (evolution). Natural historians of all kinds used to spend much of their time recording *what* persisted, classifying it and labelling it so that those answering the other questions had names to which they could refer. *Why* things persist is a question with only a few answers: phenomena persist because they are either inherently stable in an absolute sense, or because they are more stable than other phenomena with which they are somehow competing. On the whole, though, the big scientific question is always *how* things persist: what mechanism explains their existence and what common laws underlie their ability to cheat oblivion.

Ripples in water persist by means of propagation, as wave motion. It has long been realized that many other phenomena persist by similar means. A sound is a phenomenon that propagates itself in the form of waves of compression and rarefaction of the medium in which it travels. Urban myths and fashions propagate from mouth to mouth. Until quantum theory came along and complicated everything, light was visualized as a fairly simple phenomenon that traverses space as an electromagnetic wave, in much the same way that a ripple propagates across the surface of water. This is not a physics book and I am not a physicist, so I don't want to pursue the following line of reasoning too far. But if you can think of light and other subatomic particles as wave-like propagating disturbances, then the very components of which the universe is built are nothing more than persistent localized distortions of the basic fields that make up all of space. Stuff is clearly not as solid as it looks.

Intuitively, we think of the universe as if it were a painting. Space seems like a canvas on which atoms and particles are daubed like paint. This is such an ingrained notion that it is embedded in our language, and even physics texts (during the rare moments when mathematics gives way to metaphor) often talk of particles as if they were solid lumps, superimposed on the fabric of space. In truth, nobody really knows what a particle is like, but it seems that a closer analogy than a painting would be a repousée panel – a sheet of metal into which an image has been embossed. A particle or photon (an individual 'packet' of light) exists as a *disturbance* in something, rather than something superimposed on a quite separate backdrop. A wave on water is quite similar to an embossed image, since the wave is literally a dent in the water surface, albeit a moving one. A photon, a particle or a more complex atom can also be seen as an image embossed into something; this time into the electromagnetic and gravitational fields of the universe. Whereas a ripple on water propagates itself by copying the front of its disturbance into the next bunch of water molecules along, light travels in a more complex way, involving two 'surfaces' simultaneously. In my imagination I can just about picture a photon as a disturbance* in an electric field, which collapses and thus distorts the local magnetic field (the same principle that makes an electromagnet work). When it is no longer supported by the energy flux from the collapsing electric field this too starts to collapse, generating a new electrical disturbance shifted slightly forward in space. In my mind, light hops along as an alternating series of electrical and magnetic dents in space. My imagination may be way off the mark, and I have no intuition for the more bizarre ideas current among physicists, such as superstrings. But I feel fairly confident that the metaphor is not too far from the truth, at least to the extent that it is reasonable to consider matter as a disturbance of space, and not a superimposition of something onto it.

This may not seem very relevant to artificial life, but it is the first step in an argument that will show us how to create synthetic creatures and also allow us to support the notion that they are truly alive and not merely a sham. The important point is that matter and energy are essentially *made of the same non-stuff* as the rest of the universe, where 'the rest of the universe' means other forms of persistent phenomena such as genes or minds. Consequently, we can apply the same kinds of reasoning and look for the same kinds of mechanism throughout the whole of nature. This is an antidote to post-Newtonian physical science, which tries to explain the universe purely in material terms and has immense difficulties dealing with higher-order phenomena

* In *The Matter Myth* (see bibliography), Paul Davies and John Gribbin describe the soliton, a propagating, wave-like phenomenon that can persist without fading away as ordinary ripples do: 'indeed, ordinary protons, neutrons and the rest of the particle zoo can be regarded, in a certain basic sense, as solitons in the appropriate force field'.

such as those that interest us here. Descartes's dualism asserted that the universe was made up from two entirely different qualities: matter and spirit. Materialism then countered with the view that matter alone was sufficient, and so the immaterial world slowly lost its status.

Only the other day I gave a radio interview about the possibility of constructing conscious machines. To balance my views the producer brought in a well-respected professor of engineering who proffered the theory that 'consciousness may be a fundamental property of matter'. He seemed to be suggesting that consciousness might be a 'force', akin to gravity, that showed itself whenever enough small components were brought together in one place. Here we see the pernicious influence of materialism at work once again. Spirit has to be reified. All intangible things must be made tangible. Wishy-washy spiritual notions must bow down before the great god of matter. In the few seconds available to me I tried to explain that there are vast numbers of ways in which small elements can be put together, but only a very, very few of them would result in consciousness. Consciousness cannot therefore be a property of matter, only a property of certain *configurations* of matter. The purpose of this book is to discuss exactly how we might configure large numbers of small components to create life, and perhaps consciousness. Approaching the task from the point of view of a physicist is absolutely the wrong way – materialist, reductionist thinking will just not do. It is better to let go of our innate preference for substance over form and learn instead to see the material world in immaterial ways. Matter is merely a minor subset of form; hardware is simply a variety of software; there is essentially no such stuff as stuff.

In the next chapter I want to continue this thread by building a series of more and more complex persistent phenomena onto the simple ones we call matter. Before then, to make sure the point has been driven home, I shall give you an example of an artificial system that knocked me off my chair the first time I saw it, and started me thinking about the world in a completely new way.

Enchanted light bulbs

On the wall of the departures hall of SEA-TAC Airport near Seattle, there is (or was, last time I looked) a large grid of light bulbs, connected to a small console that allows passers-by to draw simple patterns in the lights. But this is no mere digital sketch pad: it is a rather large and impressive example of a game invented by the mathematician John Conway, which he called (rather fittingly) Life. In Conway's game, each light bulb (or more commonly each pixel on a computer screen) is a very simple machine. Each is able to sense whether its immediate neighbours are switched on or not, and then follow a simple set of rules accordingly. At every tick of a clock, each bulb examines the state of its neighbouring bulbs and decides whether to light up or extinguish itself according to the number of lit neighbours. If there are less than a certain number of lit bulbs adjacent to it, a light will 'starve' and become extinguished. If there is a large number it will become 'over-crowded' and also go out. Under any other circumstance, the bulb will illuminate. Once all the bulbs have had a chance to decide their fate, they all switch to their new state, and the process is repeated. If all the lights are initially off or all of them are on, then the next state will be complete blackness, because all the bulbs will starve or become over-crowded simultaneously. But if a few bulbs are manually switched on before the clock is started, some will extinguish after the first tick, while others will light up.

The results can be quite startling. Sometimes the whole pattern will flicker for a few ticks of the clock and then quickly die out. Other times, stable patterns – persistent phenomena – will emerge. Some of these are trivial and boring. For example, a group of four lit bulbs in a square will just remain in the same state for ever (or until some nearby pattern encroaches on the space and disrupts the square). Lines of three adjacent lights are slightly more interesting, because they alter-nate endlessly between two patterns: one horizontal and one vertical. Such shapes are dubbed 'spinners'. Quite a large number of initial pat-terns indulge in rather more spectacular behaviour. They transform themselves into pattern after pattern, never quite repeating them-selves, and usually growing to fill a large area of the grid. Such patterns are persistent but not immortal, and they are fascinating things to study. But the pattern that knocked me off my chair when I first saw it on a computer screen in the late 1970s is a rather simpler one (see Figure 6). It is made up of five adjacent but asymmetrical lit bulbs. Its posh name is the R-pentomino, but it is commonly known as a 'glider', because that describes exactly what it does. After the first tick of the

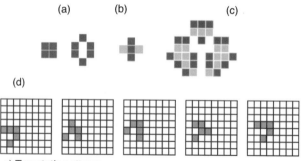

a) Two static patterns
b) A spinner (two states)
c) A long-lived, many-state pattern
d) Five steps in the life of a glider

Figure 6. Life in all its glory.

clock, the pattern changes to a slightly different one, also made from five dots. After the second tick it changes again. But by the fourth tick we find ourselves back with the original shape once more, only this time it is displaced one square diagonally from its original position. The glider continues to alternate between the four states, and with each cycle moves one square across the board. The thing that really startled me about this, and still spooks me today is that *something* is clearly moving across the board, and yet *no thing* is moving! The light bulbs don't move; they just switch on and off. There is no central controller deciding where to put the pattern next (like those displays of moving text you sometimes see in airports); it just emerges all by itself. The glider is a *thing* – a coherent persistent phenomenon that moves across 'space' – and yet it is not separate from or superimposed on that space. It is simply a self-propagating disturbance in the space created by these little rule-following light bulbs.

It seems to me that the natural world is rather like this too. Space is not really like a grid of light bulbs, and the phenomena that persist in physical space are different from those that emerge in Conway's game. But in both cases we see phenomena arising out of local disturbances, and many of those phenomena persist. We can classify them in terms of how and why they persist, and the more complex phenomena can 'hitch a ride' on simpler ones, so that every new class of phenomena opens the door to the creation of further classes. Whether we are dealing with the simplest or the most complex phenomenon, the same basic concepts and mechanisms apply – everything is a self-maintaining pattern in a sequence of cause and effect. The universe is not made of stuff but of events and relationships.

Now that we have finally put matter in its proper place, we can start to look at how whole hierarchies of these persistent phenomena, including many that have hitherto been relegated to the realms of the intangible and thus beyond the pale for science, can come into existence and persist for extended periods. Among these phenomena, besides particles and atoms, we shall discover life, intelligence and mind.

* *

CHAPTER FOUR

LEVELS OF BEING AND THE GENERAL SCHEME OF THINGS

There is nothing in the laws, concepts and formulae of physics and chemistry to explain or even only to describe such powers. 'x' [the property unique to life] is something quite new and additional, and the more deeply we contemplate it, the clearer it becomes that here we are faced with what might be called an *ontological discontinuity* or, more simply, a jump in the Level of Being.

E.F. Schumacher, *A Guide for the Perplexed*

A hundred metres below the ground, roughly between Lake Geneva and the Jura Mountains in Switzerland, sits one of the meanest, most brutal machines yet created by human artifice. In this apparatus, defenceless subatomic particles get hurled around a 27-kilometre circular tube at speeds of up to a thousand revolutions per second before being smashed head-on into one another, simply to see what happens. This is macho science at its most extreme, and reminds me strangely of a bully I once knew in school.

I recently spoke to a couple of physicists during a visit to this particle accelerator at the European Laboratory for Particle Physics, and asked them a rather naïve question, to which I didn't get a very clear answer. I wanted to know whether they thought they were smashing particles to bits to find out what they were made of, or whether they were smashing them together to see what they made. In the standard theory of matter, particles such as protons and neutrons are 'made of' smaller ones called quarks. I wanted to know whether these physicists really believed that the particles were made of quarks, or whether quarks simply represented or predicted the limited repertoire of new objects that could be made, given the energy and other quantities available in the colliding particles. I thought perhaps that when particle physicists say that they have discovered a new particle they might be using the word 'discovered' in a way that sculptors sometimes do when they say that their task is to remove the stone that is not needed, in order to discover the sculpture that was 'there all along'. As I say, I didn't really get a satisfactory answer, and I still do not really know whether physicists believe that a proton *contains* three quarks or whether it can simply be *described* in terms of quarks.

The universe never forgets a new trick

Nevertheless, it is clear that some disturbances in space and time can, under the right circumstances, give rise to other kinds of disturbance. If those other kinds of disturbance are stable, they will persist, and so the number of different sorts of pattern in the universe will increase over time as new mechanisms for persistence are 'discovered' and exploited by nature. Perhaps during the Big Bang an initially smooth universe became 'buckled' in a huge variety of ways. Because of the small number of intrinsic properties of this 'surface', some shapes and styles of disturbance were stable while others were not. Just like the domes and pillars on our Irish lake, the forms that were unstable quickly disappeared, while the ones that were stable remained. As the universe cooled, some of these particles gave rise to new ones and the repertoire of forms increased.

Different types of persistent phenomena have different properties and therefore different degrees and methods of persistence. Photons and ripples simply career around the universe or a pond at a constant rate as if they were running forward to avoid falling over. When they meet each other, the effect is something called interference: they may momentarily add to each other or cancel each other out, but this has no impact on the waves themselves; they merely pass through each other and carry on regardless.

Just suppose there are rare circumstances when two or more waves find themselves in the same place at the same time and the result *does* affect them. Let us imagine that two large waves reach a combined height that exceeds the capacity of the 'water' to support them.

Perhaps some new configuration of space arises during this encounter that happens to be stable. If so, there is no way back for the waves: the ratchet has clicked on another notch and this new phenomenon will now persist instead in their place. We can talk as if the waves have combined to create a new entity, although I don't think this quite reflects what has happened and it would certainly be wrong to say that the new entity is 'made of' waves. Of course, every time this rare event takes place and such a new phenomenon comes into existence, it remains. We could postulate that photons occasionally did this and created new kinds of particle, or that particles 'discovered' atoms. The simple palette of colours with which the universe began got mixed in various ways and new shades emerged.

Systems like ripples that persist by propagation are doomed to a fairly boring existence – they just hurtle around endlessly, for the most part ignoring one another. Atoms, on the other hand, are relatively slow-moving and localized in space, and so have time to interact with one another. Also, two atoms cannot occupy the same location, and their persistence would lead to competition for space if that space didn't happen to be infinite. As it happens, there is still plenty of room left in the universe, and isolated particles and even atoms are mobile enough and usually mutually repulsive enough to spread themselves out amicably. Atoms do interact, however, and frequently they combine to become something else that is inherently stable, which we call a molecule.

Remember that phenomena such as atoms and molecules are still only ripples on a pond; they just happen to be more complex patterns of ripples, whose mutual interactions enable them to persist. In a way, an atom is to a photon what a whirlpool is to a ripple: each has a different answer to the question of 'how' it persists, but they both share the same 'why' – they persist because they happen to be inherently stable in an absolute sense. Once certain more complex molecules start to interact, however, they can form structures in which the 'why' rule changes significantly, as we shall shortly see.

Collaboration leads to competition

Chemistry was never my favourite science when I was at school, but my memories of chemistry lessons are at least colourful. My first chemistry teacher was appropriately named Mr Cork, and he was one of those people who should never be allowed near anything sharp or expensive. He was so accident prone that we were able to revise for tests by staring at the science lab's ceiling and recalling the origins of the various stains and smudges to be found there. Unfortunately, apart from silly pranks like the day we set someone's lab coat alight and then made him laugh so that he was unable to blow out the flames, all I really remember from chemistry lessons is an assortment of minor explosions, noxious fumes and the panicked expressions of a series of long-suffering teachers.

Something I did grasp from the experience, though, was that chemistry is about *mutual* stability. Substances that are quite stable if you keep them isolated from each other may cease to persist when you bring them together. They react with each other to produce something new, and that something is more stable than the substances that gave rise to it. The more interesting reactions are accompanied by the release of a substantial amount of energy and the creation of a new stain on the chemistry lab ceiling, but sadly most of the chemistry we do in school is about steadier systems that settle down to equilibrium and stay there. Maybe that's why I found the subject dull, compared with biology.

To understand how a chemical system reaches equilibrium, imagine an infant classroom full of children inexplicably wearing odd shoes. At the start of the experiment each child is wearing either two left shoes or two right ones. Whenever a pupil with two left shoes bumps into another child wearing two right ones they swap a shoe, so that each now has a complete pair. The shoes loosely represent the components or ions of a molecule; the swapping of shoes represents a chemical reaction. As the children run around at random, the chances of a meeting between two children wearing odd shoes is initially very high, so the reaction happens quickly. After a while the probability gets lower, because more of the collisions will be between children who already have a matching pair. Eventually, though, each child will be wearing a correct pair of shoes and no further swapping will take place. The system will have reached an equilibrium in which nobody is wearing odd shoes.

We can make things a little more realistic by introducing another rule. Suppose that when two children meet and each already has a matching pair of shoes, there is a small chance that they become confused and swap, ending up with odd shoes again. All chemical reactions can happen in both directions like this, but one direction is usually much more likely than the other. Obviously this 'reverse reaction' cannot happen at all at the start of the experiment, because none of the children has a matching pair of shoes yet. But as the number of matching pairs increases, so does the rate at which this new rule applies. Meanwhile, the conditions for the first rule are being met less frequently as the number of children wearing odd shoes decreases. The two kinds of exchange will eventually reach the same rate of occurrence and balance each other. From this point on, although individual encounters between children will still sometimes involve the exchange of footwear, the proportion of children wearing odd shoes will, on average, remain the same. The classroom has reached equilibrium once more, but with a different concentration of sore feet.

A reaction that reaches equilibrium finds itself in a persistent state, albeit a fairly dull one. If we remove some of the children who have matching shoes, then the proportion with odd pairs will increase, and collisions will result in more matching pairs being created to take their place; the status quo is thus maintained. You could say that an equilibrium mixture of odd and even shoes is a persistent phenomenon in its own right. Nevertheless, far more interesting phenomena can arise in systems that never reach equilibrium.

Let us stretch our footwear analogy a bit further. Imagine our children are very young and don't really understand the concept of a matching pair of shoes. Suppose that under normal circumstances it hardly ever occurs to anyone that they should swap shoes at all. But if two children wearing odd shoes meet each other in sight of a child who already has a complete pair, perhaps this will give them a hint about what to do. In these circumstances, very few children will swap shoes initially because the idea simply doesn't occur to them. But as the number of matched pairs increases, more children will get the hint and swap their odd shoes. This system rapidly pulls itself up by its bootstraps (so to speak), because more matched pairs beget even more matched pairs. Here we have a simple case of what is known as *autocatalysis*. A catalyst is something that facilitates or speeds up a chemical reaction without being consumed in the process. For example, the catalytic converter in your car speeds up the conversion of carbon monoxide into carbon dioxide by passing the exhaust gases through a platinum mesh, which catalyses the reaction. In our experiment, the children with matching shoes are catalysing the shoe-swapping reaction. This is not just ordinary catalysis, though, but self-catalysis, or autocatalysis, because the result of the reaction that is prompted by the matching shoes is itself a pair of matching shoes, which can act as a catalyst for further reactions.

It is not outrageous to say that the population of matching pairs of shoes in this experiment is 'growing' by 'eating' all the odd shoes. It would be wrong to say that the individual matching pairs of shoes are eating or growing, but in a sense the overall process of shoe matching is a self-maintaining phenomenon. In the closed system of our classroom the phenomenon is short-lived because it will quickly consume all the available odd shoes, and the process will come to a halt at equilibrium. On the other hand, in an open system, where children with odd shoes are available in a steady and endless supply, shoe matching will be a phenomenon that persists indefinitely.

Autocatalysis of this simplicity may not seem a particularly compelling candidate for a persistent phenomenon, but it is something of a qualitative step up from atoms and molecules – a small jump in the level of being. In materialist language we would call a molecule a thing and autocatalysis a process, but I don't believe the distinction is really as fundamental as this terminology implies. Atoms are processes too, and autocatalysis is perhaps only different from atoms by virtue of the fact that one is 'made from' the other. Processes like autocatalytic chemical reactions can be seen as 'higher-order' phenomena than molecules because they are superimposed on them, but they are not fundamentally different – each is made from the same non-stuff.

The metaphor of shoe matching only gives a hint of the ideas I am trying to describe, but we have touched on several important points nonetheless. Notice, for example, that we have a phenomenon (shoe matching) that persists while its substrate (shoes) is in constant flux. In the previous chapter we looked at clouds, people and even hammers and found the same feature. We can also see how one form of persistence (such as the existence of shoes) opens the door to further possibilities – new clever tricks for existing that capitalize on the older ones. As we move from a closed, equilibrium-reaching system to an open one we also see glimpses of new possibilities. Closed systems can hold a pose well enough, but open systems can dance. We can see this with water. In the closed system of our Irish lake we recognized only two phenomena that persisted for very long: ripples and whirlpools. Rivers are open systems, on the other hand, so they can support other phenomena too. For example, in fast-flowing streams I've frequently seen an effect that looks like a piece of fishing line stretched across the surface of the water. I have no idea what causes it, but it is very persistent.

My footwear analogy didn't tell us much about the spatial nature of such phenomena. In fact, in an infinite sea of odd shoes a single, spontaneously generated matching pair would lead to a rapidly expanding 'bubble' of shoe swapping, looking something like the growth of a bacterial colony or the shock wave from a nuclear explosion. Such a phenomenon would spread like a breaking wave, with odd shoes in front of it and nothing but matched pairs behind. Where one wave had broken, no other waves could pass unless the shoes eventually became mixed up once more. Such a phenomenon might be long-lived in a global sense, but to an observer on the ground it would seem like a fleeting event because it travelled so quickly.

But such 'explosive' autocatalytic systems are not the most interesting kind. Happily, devastating single-step autocatalytic processes are rare, at least at the temperatures and pressures we are used to. Any that did exist would have burned themselves out long ago. As it happens, the probability that any one substance is a catalyst for a reaction that produces more of itself is very low. Platinum may be a good catalyst, for example, but either it cannot catalyse any reactions that produce more platinum or the appropriate platinum-bearing source substances are too rare for the reaction to reach critical mass. To find the more interesting autocatalytic systems we must look to another kind of catalyst – the enzyme.

Enzymes are proteins that accelerate quite specific chemical reactions, and these reactions are often between other proteins – including other enzymes. They are very complex molecules, and so we would not expect them to occur naturally in noticeable quantities. Yet if it were possible for an enzyme to make more of itself, it would quickly bootstrap itself into existence. To do this in a single step it would have to combine carbon, hydrogen, oxygen, nitrogen, sulphur and phosphorus together in a single reaction to produce a complex polypeptide molecule, and such a reaction is absurdly improbable. But it is conceivable that such a process could occur over many smaller steps. Many of the ingredients would be inorganic substances such as carbon dioxide and water, which are freely available. These would have to be combined through the operations of a complete network of catalytic reactions in such a way that every one of the enzymes in the network was itself a product of the system.

If such a network were possible, it seems likely that it would consist of hundreds of intermediate steps and require dozens of enzymes. It would be much less explosive in its behaviour than a simpler autocatalytic system, partly because protein chemistry is slower, and lower in energy, than inorganic chemistry, and partly because of various difficulties in keeping the much larger number of ingredients physically close enough together.

Such complex, self-maintaining networks may seem ridiculously implausible, but I know for sure that they exist in very large numbers. I know this because I am one. Every living creature is a vast autocatalytic network, taking in raw materials from outside and making more of itself, completely automatically as a consequence of the nature of its ingredients. When living things act as autocatalytic networks, we call the process *metabolism*.

Notice that the 'how' of persistence is changing as we move up the hierarchy. Free particles persist by copying their shape forward in space; atoms and molecules persist through more complex mutual interactions and resonances. Now we have autocatalytic networks, which persist by growing. These networks are not really made of atoms, or at least not the same atoms from moment to moment. They are self-maintaining patterns in space and time in their own right – persistent eddies in a flowing stream of molecules and ions.

Each of these phenomena uses a different mechanism of cause and effect to maintain itself. The 'how' of persistence has become steadily more complex at each level. Yet in the simpler cases, before the emergence of autocatalytic networks, the 'why' of persistence remained unchanged. Phenomena such as photons and atoms persist because they are inherently stable. But once we get to autocatalytic networks a whole new 'why' emerges. No longer is it enough just to be stable. Autocatalytic networks consume resources – they eat food. If this food source runs out, either because the network has depleted it or because another, nearby network has got there first, the autocatalysis stops – the network dies. In other words, in the presence of autocatalysis it is not enough for a phenomenon to be stable and simply mind its own business: *it has to be more stable than its neighbours*. If two different autocatalytic networks meet they may compete for resources, and the one that is more efficient will survive at the expense of the other. Indeed, autocatalytic networks might even eat one another.

So, as we get closer to the kinds of persistent phenomenon we might call life, the rules start to change. Absolute stability is replaced by relative stability as soon as new mechanisms of persistence such as autocatalysis come into the picture. Autocatalytic networks persist but their food molecules don't, so we may regard autocatalytic networks as a more successful mode of persistence than discrete molecules. Also, some of these networks are more successful at maintaining themselves, more *fit for survival* than others.

Absolute stability gives way to relative stability because these systems compete among themselves for space and resources. Autocatalytic networks would not threaten one another's existence if they were a long way apart or differed in their eating habits. But because the spontaneous development of a network with autocatalytic properties is a very improbable event, and because it is in the nature of autocatalysis that it makes more of itself, autocatalytic systems would tend to find themselves in close proximity and intense competition more often than not.

Obviously one cannot compete with oneself, so a single small network that consumes resources and grows into a single large network, even though it consists of trillions of molecules, could be thought of simply as a successful individual that has grown larger. It is only when parts of a network change in some way, so that slightly different persistence mechanisms are at work in different regions of space, that we can think of a competition ensuing between different individuals.

Such a change in the pattern of reactions might arise through damage to a part of the system. Proteins are molecules made from long, tangled chains of smaller components, and cosmic rays or other environmental hazards (including encounters with other enzymes) can cause proteins to change form. When they change, they take on different properties. They might no longer catalyse the same reactions as before, and the network might be weakened by every such mutation. On the other hand, a mutation might make them more efficient catalysts, or they might start to catalyse a different reaction that is also ultimately autocatalytic in the existing context. Such accidents might occasionally lead to fitter, more effective networks that persist for longer. Damage leads to differences, which lead to competitiveness.

Nobody knows whether such autocatalytic systems ever actually arose by accident, although the biologist Stuart Kauffman's work on the mathematics of such networks (see the Bibliography) suggests that they could arise more easily than it might seem. If such phenomena could arise spontaneously, they might have been the precursors of life on Earth, and this indeed is Kauffman's claim, made in his book *At Home in the Universe*, that 'Life, at its root, lies in the property of *catalytic closure* among a collection of molecular species. Alone, each molecular species is dead. Jointly, once catalytic closure among them is achieved, the collective system of molecules is alive.' This might be true, but two other vital mechanisms must also have emerged before life could get very far. Once these clever tricks came into existence, the landscape of persistent phenomena changed for ever.

Experiments you can do in the bath

The first of these clever tricks is encapsulation. In fact, wrapping up the components of an autocatalytic network into discrete packets may have been a prerequisite for their formation, because autocatalysis might not take place at all unless the components are kept at high concentration. The original containers for autocatalytic networks might have been lucky accidents: perhaps small pores in the concretions from an underwater thermal vent, or maybe natural soap bubbles. Either way, at some point autocatalytic networks discovered a way to make their own containers.

Proteins and other complex organic molecules have physical properties as well as chemical ones, and the formation of bubbles is a natural and inevitable consequence when certain kinds of organic chemicals called lipids get mixed with water. If an autocatalytic network became complex enough for these lipids (which would have been created by the chemical action of enzymes) to form a part of the web of reactions, then the whole network would tend to wrap itself up in protective, concentration-enhancing bubbles of its own making. One of the features of these lipid membranes is that they are permeable enough to allow small 'food' molecules in, without letting the larger enzymes and intermediate products out. Keeping the insides in but not all of the outsides out is an important trick, and an essential step in the evolution of life.

When bubbles get too large they burst, but often they re-form quickly into two or more smaller bubbles. So, as an autocatalytic network creates more of itself, it may grow until it reaches a point where it automatically divides into smaller units, each of which will then grow and divide again. If some serendipitous 'damage' occurs to one of the individual protein molecules, then this molecule will end up inside one of the daughter bubbles (or cells, as I think we can now describe them) but not the others. From that point on, the two daughter systems are slightly different. Each can accumulate more changes, and each will follow a different path through the map of possible autocatalytic networks. Because slightly different proteins will have different efficiencies, the bubbles will develop varying abilities to convert raw materials and so will grow at dissimilar rates. Competition for resources will ensue, and the fitter, more efficient bubbles will gradually and imperceptibly become more successful than the others. This is essentially the process of evolution by natural selection, but not quite as we know it today.

The trick is to eat the meal, not the recipe

As far as I can see, it is perfectly possible for a form of self-replicating entity to exist in the way I have just described. Bubbles containing autocatalytic enzyme networks could grow and eventually divide or bud, producing 'daughter' bubbles, each of which contains the full complement of enzymes required for autocatalysis. Any individual molecule that was accidentally mutated in the parent bubble would end up in one daughter rather than the other, which might modify the behaviour of that daughter's network and open up new possibilities for future viable mutations. Such a modified succession of bubbles could be regarded as a new variety or species of network, and might find itself in competition with the original variety. The more successful types of bubble would come to form a larger proportion of the bubble population, and some of them might lead to even more successful networks.

I think this could happen; perhaps this is even what did happen. Maybe for millions of years this new, highly successful form of persistent soap bubble gradually took over our planet. However, the development of such a phenomenon would be rare if it ever happened at all, because something important was still missing. Obviously, a mutant protein molecule in an autocatalytic set will remain part of that set only if it manages somehow to catalyse its own production. In our classroom example, a matching pair of shoes was visualized as a catalyst for making more matching pairs. Suppose that one child who originally had a matching pair of shoes accidentally lost one, and this 'mutant' child now encouraged other children to discard one of their shoes instead of swapping them (Figure 7). In this case we are lucky: the mutation is still autocatalytic, because it makes more copies of itself (more children wearing only one shoe), and so it will continue to exist. However, in order to achieve this successful mutation we had to make two exactly compatible changes at once: the child mutated into a state in which it had only one shoe, and the consequence of this mutation just happened to encourage other children to lose a shoe as well. If the same mutation instead encouraged children to steal one another's shoes, so that every encounter led to one child with four shoes and another with no shoes at all, then the one-shoe mutant 'molecule' would not be viable and would remain alone, eventually disappearing from the classroom altogether (because all children grow up and leave

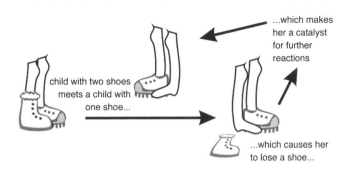

Figure 7. The one-shoe mutant.

school, and all enzymes get broken down eventually). A single altered molecule will therefore 'survive' in the population of molecules only if it changes the network in such a way that the system now produces more copies of this new molecule, without at the same time destroying the network's ability to synthesize all its other constituents.

Clearly, the chances of a mutation to a single molecule resulting in a successful new autocatalytic system, which also happens to produce more of that mutant substance, is extremely small indeed. Nonetheless, it may well have happened. In fact, it might have been a necessary prerequisite for the more sophisticated mechanism about to be described, yet it is so improbable a sequence of events that life would not have got very far on this basis alone.

However, there is a huge and quite literally vital improvement if the mechanism for carrying out the metabolism somehow becomes isolated from the mechanism that describes how to do it. Today's living organisms are certainly autocatalytic networks, but they do not rely on the network to be its own description: they store the description separately from the function, in their DNA. The DNA molecule acts as a 'template' in a sequence of events that results in the production of proteins. The structures of the enzymes that copy the DNA to make RNA, those that assemble the basic building blocks onto the RNA strand to construct a protein molecule, and those that duplicate the DNA during cell division are all specified by the pattern of the DNA itself. The whole ensemble is therefore an autocatalytic set, but with a twist.

In the autocatalytic networks we have looked at so far, a mutation would produce only a single new molecule, and however valuable that molecule might be in principle (as a protective coating, say) it would survive and accumulate enough copies to do anything useful only if it also had the remarkable ability to change the network so as to catalyse its own production. But suppose a mutation happened to the DNA instead, where the *descriptions* of the proteins are stored. The mutant DNA will now make not just a single molecule, but vast quantities of the new protein, making any improvements far more effective. Moreover, there may be no requirement for the new enzyme to take part in catalysing its own production at all. This is because the process that copies DNA frequently makes 'mistakes', and copies a portion of the molecule twice. As well as the portion of DNA (known as a gene) that produces some critical enzyme, we now have a spare stretch of

DNA that is not needed. This may make more of the same enzyme or it may make nothing at all, but either way it is free to mutate without destroying the cell's collective autocatalysis. If this spare strand eventually becomes the template for a new, useful substance, it doesn't matter in the slightest that this substance plays no direct part in the autocatalytic network. It will continue to be produced in large enough quantities to have some other beneficial effect (for example the ability to create cell membranes) and it will reliably get copied from the parent cell to any daughters. In a sense, the primary autocatalytic network, which includes the DNA protein synthesis and DNA replication functions, is acting as a host to a set of 'guest' enzymes, which tag along for the ride and presumably confer some indirect advantage on the cell, even though they play no part in its metabolism.

The finer details of how genetic encoding improves the evolvability and survivability of chemical networks are too complex to be covered here; the Bibliography suggests some further reading on this topic. The important point, though, is that we now have three vital mechanisms – autocatalysis as a means of self-assembly, encapsulation to keep parts of the network together and different kinds of network apart, and the use of templates to isolate the recipe from the function and so increase the chances of further progress. Once all three mechanisms had been 'discovered' by chemistry, a new and highly persistent phenomenon emerged on Earth (and probably elsewhere), one we call life. This phenomenon did not come into existence and then simply sit there doing nothing – it continually made more of itself, by growing and reproducing. As more and more examples of it were created, space and resources became scarce and innate stability was no longer enough. Each example of life has to compete with others for the right to exist. New clever tricks for persisting are needed urgently every day, and fortunately life has a matchless capacity for discovering them.

The snag, or perhaps the beauty of this, is that it leads naturally to an arms race. If I can't compete successfully with you for resources, I won't persist for as long. It really doesn't matter how long in absolute terms I manage to persist, provided I do it for longer than you do (strictly, provided I get the time to make more copies of myself than you do). If I threaten my own existence by rushing ahead with reproduction just in order to beat you in the race, it makes no difference that I manage to persist for only a short time – my children are multiple copies of me. The same pattern is persisting through multiple generations, rather than through self-maintenance of an individual.

Organisms can therefore trade off their life spans against their reproductive success. It is not that they have to think about this or choose to do it. Those that are optimal generators of perpetuating patterns will outpersist those that aren't. And no answer will remain optimal for very long. If I beat you by reproducing as quickly as possible, you can outwit me by reproducing more slowly but spending more effort on nurturing your offspring to increase their survival chances. If you preserve yourself by developing the ability to eat patterns like me, my descendants will grow shells to protect themselves from you. Life is an ingenious mechanism for persistence, but it is forever upping the ante by creating competition for space and resources, and in a way becomes a victim of its own success. Simple survival of the fit has to give way to survival of the fittest.

There must be more to life than this

Metabolism and reproduction are two relatively new mechanisms that the universe has discovered by which disturbances in space and time can persist. These sophisticated mechanisms build on the simpler 'mutual support' methods that maintain atoms and molecules, and the endless 'running forward to remain upright' that characterizes electromagnetic radiation. In the long, fuzzy hierarchy of forms of persistent phenomena, I think the best word to describe this region of actively self-maintaining systems is 'life'. Life is a poorly defined term but a useful one, and if it is going to be applied to anything specific, this is it. Patterns that persist by metabolizing and reproducing are alive.

But this is just a technical use of the word – an appropriate label to apply to an indistinct but important category of persistent phenomena. Just because I say that life is a good technical term for these classes of system, I don't mean to imply that life is no more than this. In common usage, life means something different – something richer and perhaps more profound. Although reproduction is all there is to life in a technical sense, everyone knows that there is more to life than sex.

And there certainly is more to life and to persistent phenomena than this. In the wake of metabolizing, self-reproducing systems came a new opportunity and a new challenge. The opportunity was the vastly increased ability to discover new persistent patterns through evolution by natural selection. The challenge was the ratchet that drives this process of evolution onward: the need to change constantly in order to stay ahead of the crowd. But evolution and metabolism are not the only solutions to the problem of how to persist in a hostile and changing world. Evolution allows offspring to be better adapted to a changing environment than their parents were, but sometimes environments change too quickly for that to be enough.

Food, for example, gets scarce in a competitive environment. Without it, the autocatalysis grinds to a halt, the pattern fails to get replicated and a previously great mechanism for persistence simply disappears. When a living network has exhausted its local food supply it can sit and wait for more to arrive, but this may not happen quickly enough. So if food molecules aren't simply going to arrive when they are needed, then Muhammad has to develop the ability to go to the mountain instead. Autocatalytic networks that can move will survive better than those that can't.

Blundering around at random on the off chance of encountering something edible will suffice up to a point, and I have to confess that this is often my own preferred method. Nevertheless, it does not take much for the concentration of available food energy to slip below the amount of energy required to find it, and so any phenomenon that can seek out and navigate efficiently towards food will have a distinct advantage over one that can't.

If your food is inanimate, you merely have to locate it and move towards it. If your food is another autocatalytic network, on the other hand, then you are a predator and it is your prey, and the chase is on. Networks may avoid being eaten by evolving hard shells or spiky skins, or by pretending to be something else, but it is better on the whole to be able to run away from predators. Similarly, if your own food starts running away, it is extremely helpful to be able to run after it.

Smarter than the average bacterium

Competition, predator–prey relationships and a rapidly changing environment favour mechanisms that can respond to the state of the world. 'Adaptive behaviour' was a new mode of persistence that became simultaneously desirable and possible almost as soon as evolving, metabolizing systems emerged in nature. Adaptive behaviour has its roots in chemical regulation. Any autocatalytic network worth its salt is likely to contain self-regulating mechanisms of some kind or another. In fact, slowing things down and controlling them is a more demanding task for a cell than it might at first appear. Without careful regulation cells would grow at an exponential rate and essentially burn themselves out. It is surely no coincidence that one of the most common symptoms of a cell whose mechanism has broken down is cancer – uncontrolled growth and cell division.

An organism's ability to adapt to changes in its environment can take many forms. Even our original population of odd and even shoes can be seen as adaptive to a degree, because it automatically responds to the addition or removal of children in such a way as to maintain its equilibrium. More sophisticated mechanisms for self-maintenance are generally based on less direct relationships between the environmental change and the appropriate response, as we shall examine in more detail when we come to design our own intelligent system.

The important point here is that adaptive behaviour can emerge from essentially the same kinds of regulatory mechanism already employed by autocatalytic networks. So this rudimentary form of intelligence can be seen as the next rung on our ladder of mechanisms for persistence – another level of being. In fact, adaptive behaviour is not quite synonymous with intelligence. Adaptation is a solution after the fact: if walking forward hurts because you have just hit a rock, stop walking or change direction; if it feels good, do more of it. Any phenomenon that can modify its own behaviour as a consequence of the environmental stresses imposed by other phenomena has a greater chance of persisting than one that just stands there and takes it on the chin, and this is what we mean by adaptation (evolution is also a form of adaptation, of course, but here we are referring to adaptation within a single lifetime).

Adaptation comes in many forms, from simply learning to ignore repetitive changes that cause no harm, through to sophisticated mechanisms that can elicit the right responses in a wide range of situations. This ability to react to events after they have happened can get an organism a long way, but not nearly as far as *pre*-adaptation can. In other words, reacting to an existing opportunity or problem is not as effective as *predicting* it and changing one's behaviour appropriately before the opportunity has time to go away or before the damage is inflicted. Intelligence is perhaps a term that should be reserved for systems that can predict the future.

Reflecting on the future

Propagation, as in a ripple or a photon, is a feed-forward mechanism. One effect becomes the cause for another, and the whole thing happens in a linear chain. It is rather like a domino run – one domino falls and triggers the next. Autocatalysis, on the other hand, is a feed-*back* system. The dominoes loop around on themselves; a reaction product becomes an ingredient in the recipe for its own creation. Adaptation is also a feedback process. Changes in the environment feed back on the organism through its senses and cause changes in its behaviour. When a creature walks into a rock, the feedback from its senses cause it to alter its behaviour and stop or turn.

Primitive forms of learning become possible when the feedback is not directly coupled to the creature's behaviour but instead is applied to some kind of memory of the event, so that instead of reacting now, after the fact, the creature modifies its future behaviour. Such a mechanism is a simple form of prediction, because it is predicting (on the basis of past experience) that in situation A, action B will be a good response. Since it is a form of prediction, it is also a primitive form of intelligence. There are far more sophisticated mechanisms for intelligent behaviour than this, but whatever kind of machinery it uses, intelligence can be seen as a new mode of persistence – something that came into being and won't go away.

At some stage in the history of persistent phenomena, something genuinely new and rather special emerged: a mechanism somehow became capable of being aware of its own existence and able to reflect on its own present, past and future. Such a capacity allows an organism to contemplate and cogitate, to rehearse and imagine, to place itself in someone else's shoes. This is self-awareness, and it too will be discussed in a later chapter. For the moment, however, use your own consciousness to reflect on this fact: mind is another persistent phenomenon – something that has come into being and now won't go away.

Mind is not different from matter and yet it is not matter, nor is it a property of matter; both matter and mind are made of the same non-stuff.

We now have quite a towering hierarchy of more and more sophisticated forms of persistence: photons, particles, atoms, molecules, auto-catalytic networks, self-reproducing systems, adaptive systems, intelligence and mind. On top of that, or somewhere to one side, we can perhaps add society as another level of being. A society is a self-sustaining emergent phenomenon that comes into existence among populations of communicating and interdependent organisms, just as an organism is an emergent phenomenon that comes into being among populations of interdependent cells. Societies regulate themselves through laws and social mores in much the same way as organisms regulate themselves through catalysis and adaptation.

Some of these persistent phenomena look like familiar machines built from smaller physical components, in the way that brains are built from neurones. Some seem to be immaterial things built from material things, such as minds growing out of the operation of brains. Others seem even less tangible and rootless, for example societies or fashions. Yet all are persistent phenomena that emerge out of other persistent phenomena.

But what does 'emergent' really mean?

* *

CHAPTER FIVE

THE IMPORTANCE OF BEING EMERGENT

'From a drop of water,' said the writer, 'a logician could infer the possibility of an Atlantic or a Niagara without having seen or heard of one or the other. So all of life is a great chain, the nature of which is known wherever we are shown a single link of it.'

Arthur Conan Doyle, *A Study in Scarlet*

My friend Owen Holland now works in the rather exotic field of artificial consciousness, but until recently his research was conducted right at the other end of the intelligence scale, where he worked with a troupe of rather house-proud robots. These small, two-wheeled automata devoted their humdrum little lives to tidying away frisbees. At the start of the day, Owen and his colleagues would scatter frisbees randomly around the robots' enclosure, and by the time the working day was over the robots would have tidied them up again into neat clusters.

This feat would be unremarkable if it weren't for the fact that these robots knew nothing at all about frisbees or clusters or what it means to tidy things up. They had not been programmed to assemble things into heaps, and didn't even have the ability to see them. The clusters emerged instead from the operation of a very simple rule that seems to bear no particular relation to the task in hand. Each robot had a metal scoop attached to its front, like a miniature bulldozer, and this was connected to a microswitch. If the robot blundered into a single frisbee it would not be heavy enough to trigger the microswitch, and so the robot would continue on regardless, pushing the frisbee in front of it. But if the same robot then collided with a second frisbee (or a wall or another robot) the total friction would be enough to trigger the microswitch, and the robot would respond by backing up a short distance and turning by a random amount before continuing on its way. A solitary frisbee would therefore be pushed around the arena until it collided with an obstruction, such as another frisbee, and then remain there as the robot backed away and trundled off in another direction. Now there would be two frisbees together, making a larger obstruction that was even more likely to intercept further frisbees until eventually all of them had been gathered together in a few large heaps.

These experiments were modelled on the behaviour of ants, who are known to collect their dead together into neat piles, or ant cemeteries. The construction of cemeteries seems like intelligent behaviour on the part of the ants, but work like Owen's shows that it can be achieved using no intelligence whatsoever, with no 'boss ant' to lead the operation and no direct communication or coordination between the insects. When a relatively complex result arises out of simple interactions between members of a population in this way, it is known as emergent behaviour.

The word 'emergence' is quite controversial in some quarters. When populations of interacting structures become arranged in certain configurations, and something new and surprising comes into existence, we call this an emergent phenomenon. Yet some scientists dispute that these phenomena are real at all, and think they are a product of our own desire to categorize things, or that they are a surprise to us only because we are not clever enough to have predicted their occurrence.

Is emergence really only ignorance?

Unexpectedness is certainly a common feature of emergence. Nobody who was given the rules for Conway's Game of Life would immediately infer the existence of the glider. But is that perhaps because we are just not clever enough to see it? If we were smarter, would the existence of the glider be immediately obvious from the rules alone, and hence no surprise? If so, would the glider no longer deserve to be called emergent? I don't think so.

The quote from Sherlock Holmes that begins this chapter suggests that we can infer the existence of something like a waterfall from any element in its chain of existence, such as a raindrop. This typically Victorian attitude, which expects logical reasoning to conquer all, is actually more hubris than fact. Once we have seen Conway's glider we can easily work out how the basic rules led to its creation, so the existence of the glider could in principle be logically inferred from the rules alone – after all, it does not arise by magic. The fact that we fail to anticipate it is partly a problem of perspective: if I peer at you through a small hole in a fence when you are some distance away, I can see you but you cannot see me. Logic suffers from a similar one-way effect. Having seen the trick we can now work out how it was done because this exercise is a narrowing-down process, aiming towards a known goal. But going the other way is much harder. Starting with knowledge of the parts, it is more difficult to work out the behaviour of the whole because we do not know where we are heading.

Yet even with enough brainpower at our disposal, there is perhaps a more fundamental reason why the concept of emergence is justified. Certainly we could, in principle, predict the existence of a glider from the rules of Life, but only by actually carrying these rules out, simulating the system in our minds or in a more concrete form such as a computer program or an exercise on paper. In other words, to see the ultimate effect of the rules of Life, we are obliged to play out the game in our heads and see what happens. Just staring at the rules will not get us there. If we can predict the behaviour of the system only by running it, then this is no kind of inference at all – it is more like turning straight to the last page of a detective novel to find out whodunnit.

So emergence is genuinely surprising, but which things are truly emergent and which are not? Is breakfast an emergent phenomenon caused by the juxtaposition of eggs and bacon? Is soccer an emergent consequence of a group of men in shorts? What is more, are emergent phenomena real things or are they just figments of our imagination? This is why I went to such lengths to disabuse you of the notion that matter is different from form. Matter is just one link in the chain of being. Atoms are no more real than societies or minds. Hardware is a subset of software. If we accept that an atom is still a real thing after all this, then we must give equal weight to something like a mind, because they are both made of the same non-stuff. Conway's glider is a real thing too. Just because it is 'immaterial' we should not regard it as any less real than a molecule, because both are self-sustaining patterns in space and time.

But we do have to be cautious about which things are really emergent phenomena, and which are artefacts that result from the way our brains perceive and classify the world. For something to be an emergent phenomenon I think it has to be persistent in its own right. For example, a business meeting is a thing in some limited sense but it does not contain the means for its own continuation, so it is not a thing in the striking way that a society is. Societies actively maintain themselves through laws, policing and wars, while business meetings simply drag on … Neither is a soccer game an emergent phenomenon, because it does not assemble or maintain itself. A soccer match is instigated from outside, and if we do not count injury time or extra time it lasts for 90 minutes regardless of what happens on the pitch. Soccer as a sport, on the other hand, emerged out of nowhere and would probably come as a complete surprise to an alien biologist, given the basic rules of human behaviour to work from. There is no external referee waiting to blow his whistle when soccer has run its time as a social phenomenon – it continues because (in some way that I have never understood) it contains a mechanism for its own survival.

Business meetings and soccer matches are imposed from the top down – they are local consequences or manifestations of a larger system – whereas societies and sports emerge from the bottom up, as a global consequence of a population of smaller components. Yet to some degree this is just a matter of perspective. Even soccer and societies are persistent only because they form part of a bigger system, like eddies in a larger stream. This is true to a greater or lesser extent of most, if not all, emergent phenomena. Really, there is only one huge emergent phenomenon in existence, called the universe. Inside this there are regions of smaller, more or less independent, interacting loops of cause and effect that we think of as things in their own right. But usually these coherent regions remain so only by interacting with their context. We can think of minds, societies, atoms or ideas as discrete things, but we must always be mindful that they form part of a larger matrix.

The interconnectedness of all things

Sometimes the subtle interplay between distant parts of this giant super-pattern can be surprising to those people who take a less holistic view of nature. If things like telepathy or precognition have any real basis – and I am not suggesting they do – then they might plausibly involve self-organizing chains of cause and effect – as if the universe as a whole were conspiring with itself to link up things that otherwise would not be linked, in the pursuance of some grand plan (or more likely a huge joke!). Maybe poetic justice is a real force in nature – it would certainly explain a lot about history if it were. Such ideas need not be so surprising. We have seen that the long-debated 'problem' of how mind has influence over matter is spurious. Since the two are not distinct, the idea that one can affect the other should not be at all difficult to accept. We find ourselves perfectly happy to believe that our minds *control* matter, though. We have thoughts, and we implement those thoughts as purposeful action, and the world changes in consequence. If some portions of this superpattern can show purpose, then perhaps others can too. Much of this is mere speculation. Yet the notion of purposeful behaviour does bring up the general issue of what we mean by 'control', so before we look at feedback systems I'd like to make a few observations about this.

Perhaps the first thing to say is pretty obvious but all too frequently forgotten: cause and effect act in webs, not chains. There is no such thing as the prime cause of any particular circumstance; every cause is the effect of at least one other cause, and usually far more than one. For the want of a nail, the shoe was lost; for the want of a shoe, the horse was lost; for the want of a horse the battle was lost. But what caused the shoe to need a new nail? Why was the army short of horses? What started the battle?

When we draw up a family tree we generally show a small number of ancestors leading to a large number of descendants. Yet of course we could equally draw it the other way around – one child has two parents, four grandparents and eight great grandparents. Similarly, individual causes lead to many effects, each of which causes other effects. But at the same time, every effect has many causes, each of which was the effect of other causes. The universe is one huge domino run, with many dominoes falling at once, and their paths criss-cross so that the passage of one domino changes the future course of many others. To create and understand life we must remember this, and not be simplistic about cause and effect.

Another mistake we often make is to think of control as something that is imposed from outside or from the top down. Control is really as much an effect as a cause. We shall come back to this when we look at free will, but it is worth bearing in mind that a general is controlled by his troops just as much as he controls them. In a feedback loop, like any loop, there is no start and no end.

This view that control is essentially synonymous with domination has also had consequences for the way people think about natural intelligence, and how neuroscientists interpret the workings of the human brain. Our metaphors for the operation of the brain are frequently drawn from the production line. We think of the brain as a glorified sausage machine, taking in information from the senses, processing it and regurgitating it in a different form, as thoughts or actions. The digital computer reinforces this idea because it is quite explicitly a machine that does to information what a sausage machine does to pork. Indeed, the brain was the original inspiration and metaphor for the development of the digital computer, and early computers were often described as 'giant brains'. Unfortunately, neuroscientists have sometimes turned this analogy on its head, and based their models of brain function on the workings of the digital computer (for example by assuming that memory is separate and distinct from processing, as it is in a computer). This makes the whole metaphor dangerously self-reinforcing.

But there is a risk that by casting the brain in an active role as the processor of passive information we may be missing the point. Not all machines work in such an assertive and forceful way. Imagine a coin-sorting machine, for example (Figure 8). Such a machine might work by allowing coins to roll down a narrow slope between two vertical sheets of board. A series of holes of increasing size is cut into one of the boards, with a pocket behind each one. When coins of various denominations are rolled down the slope, each one will slide past the smaller holes but fall through the first one large enough to accommodate it. The coins enter in a random order and come out sorted by size. Yet the machine doesn't sort the coins – the coins sort themselves.

Such a simple, static device would probably not work in practice. Perhaps the coins would twist and get stuck in the smaller holes. In any case, we get a better analogy if we allow the machine to have moving

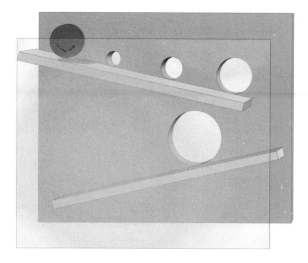

Figure 8. A hypothetical coin sorter.

parts – perhaps sprung flaps that send heavy coins one way and lighter coins another. The idea I am trying to get across is that such a machine is controlled by the coins coming into it, rather than it being the thing that controls the coins. I suspect we might get a better insight into the workings of the human brain if we take this as our metaphor, and think of the information acting upon the brain at least as much as we think of the brain acting on the information.

For example, imagine a signal from your senses entering your brain and following a certain path through all the neurones, as if it were a child running through a maze, to emerge eventually as a signal that triggers a muscle. Imagine too that the passage of each signal through the maze of neurones opens some doors and closes others, so that subsequent signals take a different route. The signals are controlling one another, using the brain as a kind of memory of their past behaviour. You shouldn't take this analogy too literally because it does have its pitfalls, but the point I'm trying to make is that control does not imply the domination of one thing over another – it is purely something that happens. The machine and the information can interact and cooperate, without either one of them being 'in control'. The image of the brain as a top-down command and control system which processes data in the way that a bureaucrat processes forms can be very misleading.

Another universe beckons

For the past few chapters I have been trying to give you a feeling for the gestalt: for form, for the whole that is greater than the sum of its parts. I have been encouraging you to take what might be an unfamiliar and perhaps disorienting view of the natural world. From this perspective, concepts that were previously clear-cut and definite are now fuzzy and intertwined. Substance has been replaced by form. Chains have turned into webs, and masters have become slaves. Rather than a fixed cast of actors we see an endless Russian doll of levels of being. Life, we discover, is a loose coalition of self-maintaining eddies in a flowing stream. When the whole becomes more than the sum of its parts something new and perfectly real comes into existence. This gestalt is not mysterious, but neither is it a figment of our imagination. It is surprising, but that does not mean it is unknowable or that it cannot be created to order. Owen Holland and his colleagues created a small emergent phenomenon deliberately and with understanding. They didn't discover rules for making frisbee cemeteries by accident – they invented them. They did this by 'constructive tinkering' – by developing a feel for how certain kinds of rules play out, and exploring these ideas in thought experiments before validating them with robots.

Frisbee-pushing is very simple, but I believe that far more complex emergent phenomena can also be a product of human art, as well as a product of accident or evolution. To achieve this we need to adopt a different mode of thought to that used by the scientist, or indeed the artist. What is required is a mindset that society has tended to undervalue in the past: the mind of the inventor. We can *engineer* the gestalt and bring new persistent phenomena into being by design. We first need to collect natural phenomena, just as Victorian naturalists collected butterflies or beetles: we must note how rivulets form in the sand on a falling tide, how water flows out of a bath, how children inevitably get themselves into trouble. We should look into the souls of these phenomena and extract their essence. But unlike the scientist, who labels it and keeps it in a glass cabinet, we must take this essence and build new things with it. We should see how the same flows of cause and effect can be used in another context to create something new – how an idea can be stolen from here, and a clever trick from there, and folded together in a stepwise synthesis that brings something glorious into being.

Soon we shall try to identify the palette of basic colours from which the rich tapestry of the universe is woven – the basic building blocks of cause and effect. This will provide us with the tools we need to create life for ourselves. But first I want to move away from our universe altogether and explore a parallel world called cyberspace, because that is where our artificial life forms are going to be created.

* *

CHAPTER SIX

LOOKING-GLASS WORLDS

Listen: there's a hell
of a good universe next door; let's go

e e cummings, *1 × 1*

Victor Frankenstein built his monster from chemicals and the vital spark of electricity. We too are going to use electricity to create life, but our electric current will animate a digital computer. Before we can understand how to create life using a computer, we must establish some important ideas about computer simulation. We need to step away from the idea of the computer as a device for carrying out instructions and instead think of it as a place in which to build other types of machine. I want to introduce the notion of different orders of virtual machine and show how this idea relates to the information in the previous chapters. First of all, though, we shall look at how a computer may be used to create virtual objects and virtual environments in the first place. How exactly are binary numbers and logical operations used to generate virtual reality?

Flying is perfectly safe as long as you keep away from the ground

It was late in the evening. Visibility was almost nil, and the windshield was awash with streaks of rain. The altimeter told me I was 2000 feet above sea level, but I must have been much less than that above the hilltops. I levelled out and eased back the speed. There was nothing to see out of the window, so I concentrated on the instruments in the cockpit. With my hand on the yoke and one eye on the artificial horizon, I tried to keep the aircraft straight and level in the heavy turbulence. With the other eye I watched the needles of the instruments. One needle told me we were on track for the runway. Another sank gradually towards the middle of the display as we approached the glideslope. As that needle centred, I extended the gear and lowered a little flap, and we began our final descent. Out of the window I could see nothing but grey as we descended through 1500 feet, then 1000 feet. How close were the tops of the hills? Where was the runway? My arms tensed and I began to sweat as the turbulence hit, making a smooth approach impossible. The needles started to drift away from their centrelines. I was down to 500 feet, and everything had gone badly wrong. The cloud started to break, but there was no sign of the airfield through the dark hole ahead of me. Finally, I caught sight of the runway approach lights, way off to my left and at an impossible angle. There was no way I could bring this beast into line now, and the ground was rushing up towards me at a terrifying pace. I quickly applied full throttle, raised the nose and turned tightly as I began a go-around. As I raised the gear and reached for my charts to set the navigation radios for a second attempt, I could feel my heart racing and my dry lips cracking. At that moment, just as a mixture of relief and embarrassment at my ineptitude flooded across my face, my wife came in and hit the pause button. Dinner was ready, and my second approach was going to have to wait.

Elegant simplicity

I have been programming computers and writing simulations for two decades, yet it still amazes me that a glorified adding machine can create an alternative reality right in front of my eyes: something real enough to make me genuinely fear for my life while I'm quite clearly sitting in a warm room, firmly attached to terra firma. How does a flight simulator create the sensation of reality, when a computer knows nothing about flying, aerodynamics or even three-dimensional space? Computers can do only a handful of things. On the level of description at which we interact with them (that is, at the programming level, rather than the logic-gate or semiconductor levels), this is really all they can manage:

Copy a binary number from one location in memory to another.
Add or subtract two numbers.
Jump to a new point in the list of instructions.
Test to see whether a number is zero, and jump to a new instruction if it is.

In practice, there are a few other things, such as the logical operations AND, OR and NOT. Nevertheless, the above are the principal actions underlying most computer code. It is reasonably easy to see how these basic operations can be combined to operate a payroll, say, or control the central heating, but how do we create virtual reality from such primitive tools?

We can illustrate the ideas by exploring a very simple computer simulation. Suppose we want to simulate a bouncing ball. This is not as exciting as simulating the flight of an aircraft, I admit, but it is much easier to explain. If you'll forgive me, I'd like to start by having a small dig at the way a mathematical education withers the brain. I have asked several university-trained programmers how to simulate the bouncing of a ball, and their logic has generally gone something like this:

'Well, a ball describes a series of parabolas as it flies. The equation for this is $y = \text{abs}(\sin x)$, but the size of these parabolas decays with each bounce, so we have to scale y as x increases. We'll also need some constants, so we have $y = \text{abs}[\sin(ax)]/bx$. Finally, friction makes the bounces get closer together, so that makes it $y = \text{abs}[\sin(ax^c)]/bx$.'

'Ah,' says I, 'but I'd also like the ball to hit a wall and bounce back.' At this point they can usually think of something more interesting to do with their lives.

This is what I call an 'outside-in' approach to computer modelling, and it is *bad*. The mistake is to start with the outward behaviour you want to see, and work back towards some equation that produces it, rather than start with the fundamental physical processes that are at work, and from them build outwards to generate the behaviour. A good deal of virtual-reality software is like this, unfortunately – it starts with how things should *look* and then tries to bolt on code for how it should behave.

Some entertainment flight simulators are written this way too, and it quickly trips the programmers up, because they are forever having to add special code to simulate stalling, spinning, and so on, instead of having these properties *emerge* naturally from a physics-based model. I used to work in a computer-games company, and I remember having a big argument about this in regard to a car racing game. The people who were writing the game had decided that, since cars describe a curve as they race around a track, the first step was to program in the shape of the curve the cars should follow. I argued that they should be trying to model Newton's basic equations of motion, and the curves would turn up all by themselves, complete with swerving, skidding and all the other things that help to convince people that they are really driving a car. Needless to say, they knew better, and they had endless trouble because of it.

Emergence is the essence of what we are trying to achieve – a thing that is really alive and really thinks, not a sham that *looks* as though it is alive and thinking. For this we must start from the inside and work out; structure must generate function. Copying the outward appearance of intelligence rather than the structures that give rise to it is cheating and doesn't really work.

Having a ball with arithmetic

But enough of the diatribe. We shall try to simulate our bouncing ball without the use of any sine functions or exponents. In fact we can do it with little more than addition. Let us begin by identifying the physical variables involved. We can represent the horizontal position of the ball by the variable X, and its height by Y. For those of you unfamiliar with computers, a variable in this context is a named memory location, like a box with a label on it, in which a numerical value may be stored. So when I say that X represents the horizontal position of the ball, I mean that a number is stored in the box labelled X, which shows how far to the right the ball is at any one moment.

At every 'tick of a clock' (in other words, every time we run the calculation) the values of X and Y will need to change by some amount to specify the ball's new position. The amount by which they change represents the ball's speed, and we can store these rates of change in two new memory locations that we can call DX and DY ('D' for 'delta', meaning 'change in'). Speed is the rate of change in position, and the rate of change of this rate of change is acceleration – in other words, it is the amount by which DX and DY change every tick. Since the accelerations in our simple model are determined by friction and gravity, and these do not change from moment to moment, we can treat them as constants and do not need to store them in memory variables.

So, imagine our ball starting in the air at the left edge of the 'world', and moving to the right as if it has just rolled off the edge of a table. Assuming our world to be 1000 units long and 1000 units high, position $(0, 0)$ being at the bottom-left, we can specify the starting conditions as follows:

```
X=0, Y=1000        ; Top-left of world
DX=10, DY=0        ; rolling right, but not yet falling
```

Again, if computer code is new to you I should explain that the text on the left is instructing the computer to do something, for example place the number 0 into memory variable X and the number 1000 into the box marked Y. The text on the right, beyond the semicolons, is a comment – an explanation of the code on the left for the benefit of the programmer or another reader. You don't need to understand any of the code in this chapter in detail. I just want you to absorb the general idea behind what is happening. What is happening in the first fragment of code is that we are telling the computer to remember the starting conditions for position and speed.

So now we have started the ball rolling, as it were, we can tell the computer to run around in a loop, executing the same series of operations over and over again. In a real simulation we would execute this loop at regular intervals by using the computer's clock, so that time did not appear to keep speeding up and slowing down as the time taken to do the calculations varied. For simplicity, however, we shall just loop around as quickly as the computer can manage it.

At each tick of the clock (or each trip around the loop) we need to add the current speed to the current position in order to calculate the ball's new position. This is simply:

```
LOOP:              ; (This label marks the top of the loop)
X=X+DX             ; Add the contents of DX to X and store result back in X
Y=Y+DY             ; Similarly, compute a new height
GOTO LOOP          ; Jump round and do it all over again
```

For every tick, the ball will now move to the right by 10 units, and fall by 0. It falls by zero units because we have not written any code to handle acceleration yet. To simulate the acceleration due to gravity we must subtract something from DY at each tick (subtract because gravity acts downwards, and we have taken upwards to be the positive direction). Because Y is changing by the value of DY at each tick, and DY is also changing, the speed of the ball now increases as it falls. We only need to add one line of code to simulate gravity, and our program now looks like this:

```
LOOP:              ; Every tick, do the following ...
X=X+DX             ; Add speed to last X,Y position
Y=Y+DY             ; to calculate new position
DY=DY-1            ; Calculate the acceleration due to gravity
GOTO LOOP          ; Jump round and do it all over again
```

If the ball hits the ground we must make it bounce, and we can easily do this by changing the sign of DY to reverse the direction of travel at the appropriate moment. Because the direction of the acceleration remains the same, the ball will now get slower as it rises, instead of accelerating as it falls. We shall know when the ball hits the ground and needs to change direction, because *y* will pass through zero.

Finally, we can add a little friction at every bounce, to slow down both the vertical and horizontal speed of the ball. The final simulation code looks like this:

```
X=0, Y=1000        ; Start at top left of world
DX=10, DY=0        ; Ball is rolling right but not yet falling

LOOP:              ; Every tick, do the following ...

X=X+DX             ; Add speed to last X,Y position
Y=Y+DY             ; to calculate ball's new position

DY=DY-1            ; Vertical speed changes due to gravity

IF Y<=0 THEN       ; If we have hit the floor ...
   Y=0             ; Don't bounce right through it!
```

```
DX=DX*0.7        ; Include some friction to slow the ball in
DY=DY*0.7        ; both the horizontal and vertical directions
                 ; by multiplying the speed by a fraction
DY=0-DY          ; Reverse the ball's vertical direction
END IF           ; That's the bounce dealt with


IF DX<-1 OR DX>1 THEN GOTO LOOP      ; Repeat until we've stopped
                                          rolling
```

Notice that we've used nothing more than addition, subtraction and multiplication – there are no sine functions or square roots. Even though this program looks longer than the mathematical equation above, it is really very much shorter and faster, since sine functions have to be converted by the computer into lengthy sequences of simpler arithmetic.

What's more, every term in our program relates directly to some physical quantity (speed, acceleration, gravity). Adding new features to our model is much more straightforward than in the more abstract, outside-in approach. For example, if we want the ball to bounce off a wall at X = 1000, we simply add the line

```
IF X>=1000 THEN DX=0–DX    ; Reverse horizontal direction
```

All that is left to do, if we are to make this into a veritable virtual-reality 'experience', is to display the ball on a screen so that we can see it bounce. In real-life virtual-reality simulations, enormously sophisticated code is used to create a display in which complex, three-dimensional shapes are portrayed, covered in realistic textures and 'lit' by virtual light sources. In our simplified simulation, everything is in two dimensions and we can treat our ball as a simple dot, so that displaying it is fairly easy.

Most computer displays use memory-mapped screens. In the simplest types the colour of each dot on the screen is controlled by the value stored in a single location in memory, and the memory addresses (the numbers that the computer uses internally to label each 'box') increase sequentially along each row of the screen in turn. To calculate which dot we need to light up in order to show the current position of the ball, we do a straightforward calculation:

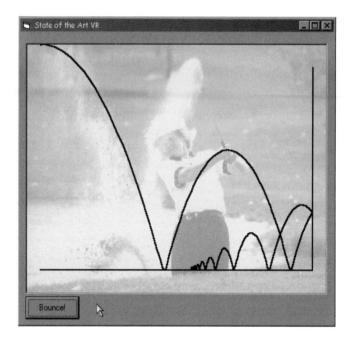

Figure 9. A virtual ball.

Dot = Addressofscreen + Y * Widthofscreen + X

At each tick, we wipe out the dot that we drew at the last tick (say by storing 0 in the appropriate screen memory location), then calculate the new X and Y for the ball's position and draw a dot at this new location by storing a '1' in the screen memory. That is all there is to it! If we do not wipe out the old dots, the track of the ball will remain on the screen and we can print the result as a still picture, which will look like Figure 9.

Forgive me for labouring this explanation of a very simple computer simulation, but for those of you with little or no programming experience I wanted to show how straightforward arithmetic can be used to create a simulation of reality; in this case a 'virtual ball'. What we have made here is a *virtual machine*, which acts like a ball. We could, with more effort, make virtual aircraft, virtual bridges or even virtual computers.

Being virtually alive

Can we simulate a virtual living thing in this way and will it really be alive? The short answer is 'no', but happily for wannabe Frankensteins, this is not the whole answer. Many people have tried to create life – or at least intelligence – by this direct approach, but I think they are wrong to do so because they are only emulating life, not creating it. The idea I want to put forward instead is this:

> A simulation of a living thing is not alive, and a simulation of intelligence is not intelligent. On the other hand, intelligent, living things can be made out of simulations.

Remember the hierarchy of levels of persistence that we looked at in previous chapters? The idea I want to explore now is that a similar, indeed essentially identical hierarchy can exist within a computer. The only place it differs from the natural one is at the first step.

Another way of explaining the above statement should make things clearer. A computer simulation of an atom is not really an atom. It does not really have mass or charge, it only *behaves* as if it does (just as the virtual ball above behaves rather like a ball, yet clearly is not one). But if you make some molecules by combining those simulated atoms, it is not the molecules' fault that their substrate is a sham; I think they will, in a very real sense, actually be molecules. If we then go on to make materials out of these molecules, it is not unreasonable to call these materials real. In fact, it is even fair to describe them as material things.

In terms of the hierarchy of organization that we have already postulated to exist in the natural world, we can restate this principle at any level we like. For example, if we simulate nerve cells using computer code, then they are not really nerve cells. But if we use these simulated nerve cells to build a brain and the brain thinks, it is not the brain's fault that its constituent neurones are a sham; it will still be a brain and its thoughts will be real thoughts. If it then goes on to proclaim itself to be conscious, who are we to deny it? On the other hand, if we simply try to write a computer program that behaves as if it is conscious, then I think it is wrong to say that the result really is conscious. I do not even think this approach can succeed, in fact, because consciousness is such a complex phenomenon that building its outward properties in a piecemeal way would be a hopelessly protracted quest, whereas if we can get consciousness to emerge by itself, we will get all of its properties for free.

I do not believe that it is especially contentious to say that a computer program that simply emulates the outward behaviour of a living, intelligent or conscious being is really nothing more than a clever fake. But my assertion that such a charade can be used to create something that is genuinely what it seems to be is far more controversial. My general argument is that a computer simulation of an entity is not in itself a real manifestation of this entity, and yet higher levels of organization built from such entities have, under the right circumstances, a genuine right to be declared examples of the things they simulate. We can therefore think in terms of different *orders* of simulation. Our simulation of an atom is a first-order simulation, but the molecules made from it are second-order simulations because they are made not by combining computer instructions but by combining simulated objects. What I am claiming is that second- and higher-order simulations are real things in a way that first-order ones are not, and that these emergent entities are freed from some of the restrictions that apply to first-order ones.

If it quacks like a duck …

That phrase 'under the right circumstances' is the bit that needs some careful qualification. There are two caveats here. The first is that we need the first-order simulation to be *sufficiently realistic*. This is something we have to consider on a case-by-case basis, and it depends on the dynamics of the hierarchy we are dealing with.

Some systems we might want to model are *brittle*. That is to say, a tiny error in the fidelity of the first-order model might be magnified out of all proportion by the time we reach the second-order level, and things will get even worse if we continue to level three and above. If we model atoms and fail to simulate their quantum behaviour sufficiently well in some crucial aspect, then the molecules we build from them will be substantially different from those found in the real world – perhaps they will bond to one another in unrealistic ways, or maybe they will not bond at all. The chemical reactions they undergo would then be very different, and by the time we tried to make something as complex as a neurone out of them the behaviour of our model would be severely at variance with reality.

If life were always like this, our aim to create higher-order models with interesting properties would still not necessarily be blocked because there may be many similar high-level structures with different but equally useful properties. It would only be a problem if our initial inaccuracy made us miss out some vital component upon which everything else depended. In the Introduction I mentioned Roger Penrose's bleak outlook for artificial intelligence, which is based on such a claimed occurrence. If some quantum property is crucial for the operation of brains and if it is impossible for such a property to be programmed into a computer, then, he argues, computers can never create intelligence. In fact I think Penrose's argument fails, mostly because he has a naïve (but by no means uncommon) conception of AI. I agree with him that no ordinary first-order computer program can be intelligent, but unlike him I think that higher-order structures can.

Not all systems are brittle, happily. In many systems the exact details turn out not to be important, and much the same behaviour emerges out of the second level, regardless of the exact implementation of the first. Many A-life experiments show this to be more common than we might expect – the same kinds of emergent phenomena can arise in systems built from very different substrates. Stuart Kauffman's work on autocatalytic networks, for example, demonstrates that they will emerge readily and spontaneously from arbitrary combinations of simulated enzymes, once the number of interconnections becomes large enough. The networks themselves may be different each time the experiment is run, but their overall properties remain remarkably consistent.

Systems that show the same kinds of global behaviour despite small differences in their internal structure may be described as *robust*, and my feeling is that robustness is characteristic of the phenomena we are interested in here. This is very important in practice because, when it comes to the brain, we have barely a clue how to build one. It is all very well concluding that a brain will be a real brain, even if it is made out of simulated neurones, but this assumes that we actually know *how* to make such a brain out of neurones. Unfortunately we don't, and that would leave us quite stuck if it were not for the assertion that brains fall into the 'robust' category. We don't have to know exactly how a brain works to conclude that neurones are good starting materials for making brain-like machines.

We can tell this from the natural world, where most intelligent systems (probably all animal ones, at least) consist of neurones, yet it is extremely misleading to talk about 'the brain' as if there was only one design. Rabbit brains are noticeably different from ours; squid brains dramatically so. Nevertheless, they are all built from the same kinds of building block, and they all show intelligent behaviour.

Imagine a hypothetical museum of technology, filled with one example of every possible machine that could ever be built by humans or evolved by nature. Some fastidious curator has arranged them in such a way that machines of similar construction are placed nearest one another. In this Aladdin's cave there may be many machines whose properties could be regarded as intelligent (or alive, or conscious). But there are an enormous number of possible machines, and we have no idea whereabouts in the museum the intelligent ones are to be found. Intelligence may be a property of many different kinds of machine, and we don't know where to look to find most of them. Yet we know that there is a broad region of natural intelligent machines based around neurone building blocks, and if we want to create artificially intelligent machines then, instead of making wild guesses, it strikes me that this is a good place to start.

Stepping into a parallel universe

The second issue raised by the phrase 'under the right circumstances' is the question of *where* these second- and higher-order constructs exist. A ball is really a ball only when it has room to bounce around. A simulated atom can express the attribute of weight only if there is some external (or internal) property called gravity, and some notion of space, within which it can move. A computer program does not contain space. Space is a meaningless notion in the context of computer code, unless you are talking about 'how long' a piece of code is, in terms of the number of instructions it contains. But everything we see in the natural world takes place in an arena that we call 'space', and if we want to create equivalent hierarchies of organization inside a computer, we have to create a virtual space of some kind in which to put them.

If you think about this really hard, it is just possible to conceive of how space might emerge out of the system itself. Perhaps in the real universe this is what happened. Perhaps space is a consequence of the universe and not simply a pre-existing container for it. As I understand it, this is what many cosmologists currently believe, but I am not familiar enough with physics to grasp how this is supposed to work. From a practical point of view, though, we can safely assume that the concept of space and many of its associated environmental properties, like friction and gravity, need to be created separately and explicitly in our simulation. It does not matter if these things are written as mere first-order constructs – simple code that behaves *as if* it were gravity, or simple geometry calculations that make it appear *as if* our virtual objects have position and motion. We are not interested in creating real gravity, only real life. So the environment we need to construct is essentially only an elaborate version of the kind of environment we gave our virtual ball.

Given some code and some data structures that behave as if the program contained real space, our second- and higher-order constructs will then exist within that space. The first-order simulation does not do this; it is still just computer code and inhabits what might be called 'procedural space', if it has to be called anything. So when we build second-order virtual entities out of first-order, simulated building blocks, we break through from procedural space into a parallel universe – a looking-glass world. From that moment on, the things we build are more or less equivalent to the things we find in the natural world, and I think they are just as real. They seem insubstantial to us because they inhabit a different universe to us and we cannot touch them. A television image seems insubstantial to us because we are separated from the location it depicts, and for similar reasons we feel as if our virtual reality is not actually a real place. Yet I suggest that it is. If we were able to cross the barrier into virtual space, it would be no less real to us. In a moment, this is what we shall imagine we are able to do.

First, though, it seems right to give this space a name, and the obvious name to give it is 'cyberspace'. Cyberspace is more commonly used to mean a rather notional, vague kind of space – the space in which two people meet when they are engaged in an email conversation, or the space where the World Wide Web is stored. Normally it isn't used to mean Euclidean (ordinary three-dimensional) space, except when referring to 3D, graphical 'chat rooms' on the Internet or virtual-reality games. However, because of the broad correspondence of ideas and, not least, because of the prefix 'cyber-', which brings to mind cybernetics, I shall call this looking-glass world cyberspace.

To recap: we have looked at how a computer program can simulate reality, at least to a limited extent, but these direct, first-order simulations are only pale imitations of reality. This is particularly important when the reality we wish to simulate is alive, because living things are vastly more complex in their behaviour than bouncing balls or light aircraft. Yet under the right conditions of sufficient fidelity (or sufficient robustness) and an appropriate spatial environment in which their properties may become manifest, we can start to assemble the virtual 'objects' defined by first-order simulations into aggregate structures inside a new universe called cyberspace. I believe that our only real hope of creating intelligent and ultimately conscious living creatures is to think in this way, and separate the task into two very different aspects. At one level we need to be computer programmers, putting together sequences of instructions. But then we must don a hard hat and become biological engineers, working with objects and combining them to make new structures.

* *

CHAPTER SEVEN

THEY CALL ME LEGION; FOR I AM MANY

Life is just one damned thing after another.

Elbert Hubbard, *Philistine*

Imagine that in the corner of your room sits the biggest, fastest, digital computer ever built. In the middle of it is a video screen and attached to it by a fat optical cable is a device looking for all the world like a ray gun. The 'gun' is really a scanner, and it has the ability to record, with only minimal disturbance, the location and properties of every atom that its beam passes through.

The computer is programmed to simulate atomic behaviour on a massive scale. Its memory contains a huge list describing the position, charge, mass, motion and assorted other parameters for trillions upon trillions of simulated atoms, and at every tick of a clock the program's equations compute the new state of each atom by taking into account any interatomic collisions or other interactions that took place on the previous tick. At the moment, the computer's data memory is filled with zeros and the program has no work to do. But as you pick up the scanner gun and start waving it around, the data describing the atoms that intersect its beam are computed and stored in the database.

In seconds, the whole of your surroundings are scanned into the machine and on the video screen you can see what looks like a mirror image of the room in which you stand. You carefully chose not to scan the computer itself, to avoid the messy recursion that would cause, but apart from that there is only one other object in the real room that has not yet been copied into the computer's database. You glance at the newly constructed virtual world and see how the simulated atoms of the virtual clock on the wall are causing it to tick in a perfectly realistic way. From this you conclude that everything has been scanned satisfactorily and the model is working. Then you take the final step: slowly and carefully, you turn the gun upon yourself.

What happens next is truly startling – to both of you. That is to say, your whole existence has suddenly and completely bifurcated. There are now two of you, and each has a different conception of what has happened. Initially, neither of you realizes that anything has happened at all. Nothing seems to have changed. Then one of you notices that the computer seems to have disappeared.

It takes a while for the significance of this to sink in: you must have been copied into the machine. Everything seems perfectly normal – you pinch yourself and it hurts, exactly as you would expect. But the computer is gone and, since that was the only object you didn't scan, this must mean that you are now living in the virtual world, processed by the computer. Meanwhile, the other you – the original – is feeling rather disappointed. The expected transformation has not taken place and you find that you are still very much stuck in the real world. But wait – what is that? On the video screen in front of you, you can see a person. It is you, and you're looking pretty shocked about something.

Leaving aside Heisenberg

Such a scanner gun is almost certainly impossible, even in principle. Measurements always disturb the system that they are trying to measure. In our case we would have to measure the position and all the other properties of each atom by bouncing something off it, and each such collision would leave the atoms of the room and your brain in a new, unpredictable state. Like a long-exposure photograph of a moving subject, the resulting data would be blurred and unrecognizable. The scanner is therefore an impossibility, but the existence of such a computer is merely wildly improbable. This machine would require an unimaginably vast memory to store all the information, and it could achieve the necessary speed only by executing trillions of operations simultaneously. Nevertheless, there are no reasons in principle why such a device could not be made, at least as far as the conclusions of this thought experiment are concerned. If we had access to the data we could, in theory at least, carry out the computations. Since this is the part of the experiment from which I want to draw conclusions, it really doesn't matter that gaining access to such data is completely impossible.

If the atomic model we have programmed into the computer is sufficiently accurate, then the data that we hypothetically copied from the real world should continue to function inside the simulation in atom-like ways. I say 'atom-like' for good reasons. Given a totally accurate atomic model and equally accurate data, the modelled and the real world should remain perpetually locked in step with each other. Unfortunately, both quantum theory and chaos theory tell us that we cannot have such accuracy. Quantum uncertainty would cause the model to diverge from reality because there is no reason to suppose that the gods of the physical and virtual worlds would get exactly the same numbers whenever they cast their dice. Chaos theory defeats us too because we know that a digital computer cannot store values with infinite precision, and the famous butterfly effect shows how even the tiniest initial inaccuracies can amplify extremely quickly as the calculations proceed.

Anyway, our thought experiment lacked total fidelity in the data because the computer was not copied into the virtual world database. If it had been (and ignoring the 'hall of mirrors' infinite regress that this would cause), then the two copies of you would have received exactly identical sensory data. But because the computer was missing in one case, this version of you received a different impression of events (the apparent sudden disappearance of the computer) and so simulation and reality diverged.

Nevertheless, although the computer model would fail to stay in exact step with reality, the simulation would remain credible. Its future history would deviate from that of the real world, but the model would still behave reasonably, just as a real atom or a real person would in such circumstances. Quantum uncertainty and chaos would not cause the model to behave irrationally, just differently. The important thing to appreciate is that the virtual copy of you would continue to believe it was you. It would share the same memories and it would feel that its life had remained continuous throughout. Not only that, but it would insist that it was conscious, and who am I to disagree with it?

We cannot really scan our atoms in the way I have described, but if we could, then in principle we could use the data to feed such a simulation of atomic behaviour. This simulation would not be accurate but it could be realistic. Part of the data in the simulation would behave like a human being, and it would have a perfect right to be regarded as conscious. Artificial consciousness is therefore not impossible. We may not have access to a scanner that can provide the data by copying an existing conscious being, but at least we know that it is theoretically permissible for us to aspire to create such data for ourselves.

Programming by numbers

Here we have an example of what I have called a second-order simulation. The first-order simulation (the computer program) merely simulates atomic motion. It knows nothing about walls or clocks or human beings. It contains no code for simulating life, consciousness or intelligence whatsoever. All these things emerge from the *data* (the positions and types of the atoms), not the code.

Computer science has historically been based on the assumption that code drives data – that a program is a series of instructions which carry out operations on numbers or other symbols, transforming them and copying them from place to place. This is the sausage machine approach I mentioned in Chapter 5, in relation to theories of the brain. Yet our thought experiment turns this idea completely on its head. Instead of the code driving the data, the data in our simulation drive the code, echoing the coin-sorting analogy we met earlier. Provide the model with data about the atoms in a clock and it will tick, but provide it with data about the atoms in a brain and it will think instead.

Something important has happened. It might just be a shift of metaphor, or perhaps a shift of paradigm, but it might be more profound than that. Computers are quite clearly processing machines, and programming is undoubtedly the act of specifying a process for a computer to follow. Yet suddenly we are no longer talking in the language of processes at all, but in the language of structures. Instead of telling the computer what to *do*, we are telling it what to *be*. In this thought experiment we have used data to tell the computer to be a room containing a clock and a human being, but what the computer program is doing is merely simulating atoms. Later, when I come to talk about 'assembling' neurones and biochemicals to 'make' a living thing, I shall be talking about programming in this new sense. The neurones and chemicals themselves will be simulations – computer code similar in essence to the code we used earlier for simulating a ball. When we talk of building things from these simulated objects we are no longer writing program instructions to tell the computer what to do, but instead we are supplying data to tell it what to be.

Stuff they don't tell you on *Star Trek*

There is one other notable outcome of this thought experiment: it demonstrates something about cyberspace. Notice that you did not *move* into the virtual world, you were *copied* there. This is because of a fundamental difference between real space and cyberspace. In real space, perhaps the most elemental physical law (aside from the tautology of persistence described in Chapter 3) is the conservation law known as the first law of thermodynamics. It states that energy (and hence matter too) can neither be created nor destroyed, merely changed from one form to another. A different way of putting this is that things can be moved, but they cannot be freely duplicated – creating two from one would violate the conservation principle by creating something from nothing.

In cyberspace this law is reversed: information can be copied but it cannot be moved. This echoes the wit in Oscar Wilde's comment that, 'the only thing to do with good advice is pass it on; it is never of any use to oneself'. Programmers may loosely speak of 'moving' a number from one memory location to another, but they really mean that the number is copied. When a computer instruction says something like 'move A to B', the contents of memory location A are copied into location B, overwriting the number previously stored at B, but A remains unchanged. To simulate the movement of a number from one location to another without appearing to duplicate it, the programmer would first have to copy the value into the destination and then destroy the original (something like 'move A to B then move 0 to A').

When people step into the transporter on *Star Trek* to have their bodies 'converted into information' and transmitted to another place to be reassembled, the operators forget to mention the bit in the small print about destroying the original. If transporters were possible at all, they would result in a duplication of the objects placed in them, not a movement of them. To balance the books something unpleasant would have to be done to the poor soul from whom the copy was made.

Star Trek aside, this failure of the conservation principle is one of the biggest headaches when creating cyberspace for living things to inhabit. In the real world, the conservation of mass and energy provides the stress that drives all of biology. In real space you cannot create something from nothing, and thanks to the second law of thermodynamics (which states essentially that energy has a tendency to disperse) you cannot even expect to hold on to what you have for very long. Self-maintenance therefore requires energy, and from this fact stems all of the richness of biological persistent phenomena. In virtual space you have to impose these constraints artificially because they do not naturally exist. However, this is not hard to do, and it does not invalidate the enterprise in any way. I simply mention it because it is important – cyberspace does not come provided with a conservation law, and you do have to implement one if you intend to create artificial life because all of biology arises from the fact that there is no such thing as a free lunch. Without the need for food, metabolism is pointless; without the difficulty of obtaining it (or some similar stress), intelligence will not emerge.

Parallel paradoxes

If I were looking for a single objection to my own arguments so far, I would say, 'You appear to be claiming that algorithms cannot be intelligent, but that these second-order structures can. Surely the second-order structures are also made from algorithms, and so are subject to the same constraints?' This is a very reasonable protest, so I shall do what I can to clarify my position. (If this apparent contradiction does not trouble you, feel free to skip straight to the next chapter, which has nothing to do with computer science.)

Perhaps the essential difference between an algorithm and a second-order virtual machine is that we can put many of these virtual machines together to build organizations. Instead of a single, serial computer we now have a group of virtual objects (for example, simulated neurones), which still do their own internal processing serially but which can also communicate with one another to produce what is effectively a parallel computation. While an algorithm simply performs one step after another, in a collection of virtual objects there are many interrelationships that exist simultaneously. Two questions arise from this. I think that second-order structures can be intelligent, while first-order ones cannot. Does this mean that intelligence is necessarily a parallel process? What is more, how can second-order machines transcend the limits of serial computation when they are clearly implemented using a serial computer? Let me begin by addressing the second question, and then return to the first one in the next section but one.

It is indeed paradoxical (to put it mildly) to claim that groups of objects acting in parallel can do things that serial computers cannot, even when those parallel structures exist within a serial computer. This cannot logically be true. Since the parallel processes are being simulated in a serial machine, the function they perform collectively must ultimately be describable in terms of a serial process. If pushed, I have to admit that what I am really saying is not that such pseudo-parallel machines can do things *in principle* that serial machines cannot, but that the best way to solve certain classes of problem is to treat them *as if* a population of parallel machines were computing them. This is a much milder argument, but it is still crucially important.

A spreadsheet program is a good example. This is a simulated parallel machine – each cell in the sheet is an independent numerical calculator, which takes inputs from other cells. The spreadsheet program does its best to treat these cells as if they were operating concurrently, in parallel with one another. Nobody would claim that a spreadsheet program can actually do things that a serial computer using conventional programming languages is unable to do. Obviously this cannot be true, because all spreadsheet programs are themselves written using conventional programming languages. Yet spreadsheets are clearly far better ways of visualizing and implementing certain kinds of operation.

Company accounts are a simple case in point. It is perfectly possible to write a conventional, serial program to calculate a company's balance sheet, but it is quicker and easier to do it using a simulation of a parallel machine, because accounting involves parallel transactions.

In fact, it is quite likely that any conventional program written to handle general accounting problems will end up looking and behaving remarkably like a spreadsheet anyway. There are a vast number of different possible programs one can write, but perhaps only those in a certain class are well suited to accountancy problems, and such programs invariably start out by simulating the ability to handle concurrent processes.

So *that's* why Old Father Time carries a scythe

Spreadsheets, and programs such as multitasking operating systems (which appear to allow you to run several programs at once), are not really parallel computers – they just make a reasonable job of impersonating them. They invariably do this by following the same principle: they simulate the behaviour of a parallel process by slicing time into discrete units. Remember John Conway's Game of Life? Each little light bulb behaves like an independent machine, acting in parallel with all the others, whereas really all the bulbs are being updated one after another by a single computer. The trick is to do all the updates within a single tick of the clock. Time is deemed to have halted while all the separate light bulbs are examined and updated, and then the clock ticks on one step. Even though each light bulb is actually updated at a different moment, as far as the simulation is concerned they are all updated simultaneously because the clock is not allowed to move on until all the bulbs have had a chance to work out their next state.

Two computing tricks are used to pull this off. One is to use a time-slicing loop, in which each program element in turn is given a slice of computer time to do its calculations. The other makes use of what is called buffered storage to synchronize the updates. If each bulb in the Game of Life were allowed to change to its next state immediately after the computer had calculated it, then the behaviour of the system would depend on the order in which the bulbs were processed. Bulbs one and three might have different impressions about the state of bulb two, because bulb two may have changed state after bulb one has been processed, but before bulb three gets the chance. To solve this problem, each bulb is allowed to work out what its next state *will be*, and remember this in a temporary memory store. It isn't actually allowed to change its state, however, until all the bulbs have done their calculations. Then all of them are allowed to change to the new state before the next update sequence is started.

By separating the 'test' phase from the 'change state' phase using buffered storage, and by looping through each element in turn with a time-slicing loop, as if it had the computer all to itself, a single computer can be made to appear as if it were really many computers, one devoted to each element in the program. Sometimes this can go wrong. For example, spreadsheets have difficulty with self-referential expressions. If cell A1 uses the value computed by cell A2, and this in turn uses the value in cell A1 to perform its own calculations, then each cell's value depends on the other, and the two cells will chase each other's tails. As long as the spreadsheet is allowed to recalculate itself indefinitely then the system will behave roughly as one might expect. For example, the values in both cells might rise rapidly towards infinity. Most spreadsheet programs run their calculations for a finite number of ticks before halting to display the result, specifically to allow time for self-referential expressions to settle down.

For the kinds of calculation normally performed in a spreadsheet this is enough, but for simulating the behaviour of many real-time processes we must recalculate the values endlessly. Whether the result is a reasonable approximation to reality depends largely on how rapidly the clock ticks and the recalculations are performed. For models of many physical systems, such as feedback loops in electronic circuits, the real system settles down to a steady state almost instantly, whereas the simulation of the same circuit may take a long time before it approximates the final value. It may also show behavioural artefacts, caused by the artificial dividing up of time into discrete units or the fact that digital computers cannot store values with perfect accuracy. Nevertheless, on the whole, most continuously varying, parallel physical systems (from the beating of a heart to the flow of traffic at an intersection) can be sufficiently well approximated using time-sliced, serial computer programs. Anyone who highlights these minor difficulties in an attempt to 'prove' that machines cannot be intelligent (and some people do) is really clutching at straws.

Just because a parrot can talk, that doesn't make it an orator

So even if parallel processes are going to end up being simulated on a serial computer, there are still reasons to suppose that thinking in parallel ways (in other words, in terms of interactions within a population) may be the most appropriate way to tackle certain classes of problem. But is intelligence one such class of problem?

For fifty years, AI researchers have been seeking to develop serial algorithms that demonstrate the ability to think. To all intents and purposes, they have failed. To make defeat look more like a strategic withdrawal, 'intelligence' is a term that has been debased in computing, and tends to be applied to any program that performs some non-obvious computation. But most of these so-called intelligent programs are really not intelligent at all. As I have mentioned before, in many cases what these programs do is encapsulate a portion of the stored intelligence of their designer. An 'expert system' designed for medical diagnosis might look very impressive in action, but it is really only applying explicit rules that were previously programmed into it by one or more expert doctors. Describing such a program as intelligent is tantamount to describing a CD player as an excellent musician because it can play guitar just like Mark Knopfler.

Intelligence is not the ability to follow rules – it is the ability to develop the rules in the first place. Interestingly, AI programs that get anywhere near capturing the true nature of intelligence usually have a structure that appears remarkably like a simulation of a parallel system. Is intelligence therefore necessarily a parallel process? If it is, then AI researchers are wasting a lot of time searching through all possible algorithms for the clever trick that is finally going to work. If intelligence is parallel, then only pseudo-parallel algorithms need to be searched. These still probably account for the majority of all possible algorithms, but at least we do have some clues about where to look.

In nature, of course, all intelligent systems are also parallel systems. Even though we feel in our heads like we are one person, thinking one thought at a time, we are really a loose confederacy of a hundred billion tiny, very stupid machines. Our brains are made from vast numbers of neurones, each operating in parallel. There is no central controller and no serial, stepwise process of execution. Intelligence is the result of billions of unintelligent processes operating concurrently.

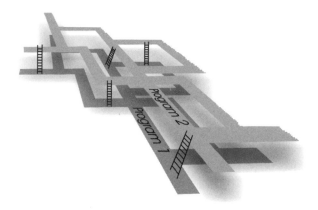

Figure 10. Parallel communicating programs.

So in practice intelligence is quite emphatically a massively parallel construct. But does it have to be, in principle? I think it does.

The topology of mind-space

One way of looking at this issue might be to convert the problem into a topological one. What is the fundamental difference between serial and parallel anyway? Perhaps the answer lies in their differing dimensionality. A list of instructions is a one-dimensional thing, like a line. You can move forwards and maybe backwards along the list, but there is no such thing as sideways. A computer program is just a list of instructions, but with a special extra ingredient because computer programs can contain 'jumps' – instructions that cause the computer to skip from one place in the list to another distant one (Figure 10). A topologist might represent this as a two-dimensional concept, because a jump is 'moving sideways', like a car overtaking a queue of traffic. It is not a true two-dimensional surface, because you cannot move around anywhere you like. Jumps exist only in specific places, so the topology of a computer program is rather more like a braid or a net. There are also conditional jumps, in which you can only go a certain way if specified conditions have been met, and jumps can even move around the code by themselves (something called a 'computed goto', in which the destination of a jump is decided while the program is running), so even a braid isn't quite the right analogy. Perhaps a sort of Möbius strip full of 'strange loops' is closer to the truth – the sort of thing Douglas Hofstadter tangled with in his famous book *Gödel, Escher, Bach*.

What now happens to a computer program if parallel processes are admitted? If these processes are simple, independent threads of execution, then they just form a set of braids sitting side by side. However, as soon as they are allowed to communicate with one another by passing messages back and forth, the whole thing takes on a kind of three-dimensional lattice structure. If a single computer program appears like a two-dimensional road network, then a group of interacting programs would look more like Boston, with freeways traversing over and under each other. In short, a parallel system is topologically three-dimensional.

This is all a bit hard for my brain to cope with, but I think it might have some relevance to how my brain works at all and why it needs to be a parallel system. The brain as a physical thing is clearly a three-dimensional structure, but is this just because of the way it is made, or are all three physical dimensions necessary for its function? It stands to reason that you could not represent a brain physically as a two-dimensional structure, because two-dimensional networks cannot have crossing paths, and most of the neuronal signals would not be able to reach their destination. You only have to draw a small set of dots on a piece of paper and try connecting them all up without the lines crossing to see how things get out of hand very quickly. Even so, we are perhaps mixing metaphors rather too much (and certainly violating the laws of syllogism) by saying that since brains are three-dimensional and brains are intelligent, then only parallel (three-dimensional lattice topology) processes can be intelligent.

There is certainly a sense in which the 'extra dimension' feels important, though. If you have ever spent any time writing image recognition software for computer vision systems (and who hasn't?), you will have experienced the sense of frustration that comes from our ability to view in three dimensions what to a computer is only one-dimensional data. The only way a computer can view a scene from a camera is to present it as an array of dots, and all arrays are one-dimensional lists of numbers to a computer, even if most computer languages allow us to view them folded up into higher-dimensional forms such as tables.

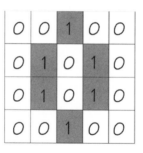

Figure 11. A digitized letter 'O'.

If you look at such a stream of dots in the only way that a computer can (one dot at a time), you very quickly see how much harder the computer's task is than ours. If I spell out the sequence 00100010100101000100, it does not immediately suggest the shape of the letter 'O' in the same way that it does when you are able to *look down on it* from a higher dimension, like Figure 11. We might think that computer vision systems are pretty stupid until we try to do the task ourselves in the way that a computer has to do it.

Are flatlanders intelligent?

Does this mean, perhaps, that our three-dimensional mental topology is necessary for our essentially two-dimensional sensory window onto the three-dimensional world to make sense? If so, I suppose it should be possible for something akin to intelligence to exist in fewer dimensions, but only as long as the organism's environment was inherently only one- or two-dimensional. Yet would we recognize this as intelligence? We tend to count a system as intelligent only if it solves problems that matter to us in our three-dimensional world. I cannot even guess what a one-dimensional survival problem would look like.

Nevertheless, it seems likely to me that there is something essentially parallel about intelligence, at least in our three-dimensional world. Probably the key idea is *simultaneity*. Turning a serial algorithm into a parallel one (even a pseudo-parallel one, simulated on a serial computer) allows several things to happen simultaneously. Various mental phenomena, however serial and unified our stream of consciousness may feel introspectively, seem to imply the need for simultaneity and therefore parallelism. When we remember things, for example, we bring previously separate things together (literally 're-membering' them), and for them to exist together they have to be represented simultaneously, side by side in parallel.

And then there is the notion of emergence. In computing terms, I imagine the opposite of emergent is pre-programmed, since something that is pre-programmed into a system is explicit within it and describable entirely in terms of the language of its parts. An emergent phenomenon is not immanent in its substrate and cannot be described in the same language as its parts. If a certain behaviour is clearly pre-programmed into a system, we would not tend to regard it as intelligent. Emergence absolutely requires parallelism because it arises out of populations of things acting in concert. Indeed, 'concert' is an apt metaphor. How much more there is to music when many voices and instruments play together than if they play one after another.

This is all rather difficult to think about, but to summarize the basic message: I do have a strong hunch that intelligence is necessarily parallel in nature and yet, as long as a computer program runs quickly enough in relation to the speed at which the outside world is changing, it is acceptable to simulate this parallelism by time-slicing on a serial computer. But perhaps there is no need for me to justify this view. If there is any doubt, it is clearly computer science that has got it wrong. After all, *everything else* in the universe is parallel: history is not one story but many; cause and effect do not form chains but networks. Computer programming is a very odd and unusual sort of thing.

* *

CHAPTER EIGHT

ON THE BALANCE OF NATURE

A party of order or stability, and a party of progress or reform, are both necessary elements of a healthy state of political life.

John Stuart Mill, *On Liberty*

Today I am recovering from a bout of flu, and for that reason I am actually sitting out in my garden enjoying it, instead of hacking at it with chainsaws and hedge clippers. This makes a pleasant change because it allows me, in between headaches, to sit and contemplate the balance of nature.

Floating in the air in front of me are several small, green larvae. I have no idea what species they belong to, but they are fascinating to behold. Each tiny creature is floating on a gossamer thread, buoyed up by the minute air currents in this sheltered part of the garden. They seem able to stay afloat for many minutes, and the conditions today must be perfect for them. If it were raining (as it usually is), they would be dashed to the ground. If the wind were stronger (as it usually is), they would become wrapped up in their own gossamer and would be unable to float along so gently and so far. But conditions are just right, and today is their day.

The thought that has been occupying my still virus-befuddled mind is how they manage to stay aloft for such a long time, and I think I now have the answer. I have just discovered that when I grab hold of their gossamer thread, they very quickly reel it in and clamber up onto my finger. If I shake them off, within an instant the thread is spun out once more and they float off harmlessly. My theory is that they are actively controlling the length of the thread, to take advantage of the ups and downs of the summer breeze. I cannot work out exactly what rule they are following, but they seem to be controlling their flight with all the skill of a modern parachutist or hang-glider pilot. I guess that they extend and retract their gossamer thread in response to the level of strain felt by some part of their spinning apparatus, and that this gives them a measure of the acceleration of their bodies and thus of the amount of lift acting on their parachute thread. When I grab hold of the top of their gossamer strand their tiny bodies are jerked to a halt, and this deceleration is a signal to them to wind in their thread very quickly. That is why they suddenly clamber up onto my finger. Shaking them loose puts them back into free fall, and so out comes the silken thread once more, and they float gently away. These tiny creatures are not simply carted around on the wind like freight dangling from a parachute – they are active pilots, in charge of their own destiny.

Just beyond the lazy aerobatics of the little green larvae, a more violent but equally controlled aerial display is taking place. Compared to the frail gossamer gliders, a swarm of flies looks like a dogfight between jet fighters. Yet these flies are just as delicately poised, and perhaps even more skilled at maintaining the appropriate conditions for their survival. I guess these creatures must be mating, because every now and again, two flies get very close to one another and a small squabble erupts. All the flies in the area burst out from the centre like children guiltily fleeing the scene of a dropped vase. Within a fraction of a second, though, they make a sharp turn and spiral back in towards the centre of the swarm.

The swarm itself has a clear boundary. Whenever a fly meets this invisible border, it adjusts its direction of flight and swings back into the rapidly rotating mêlée. The centre of the swarm remains poised in perfect balance, despite the hectic antics of its constituent parts. The entire squadron is positioned over a small column of warm air, rising from a sun-warmed paving slab. Presumably the flies can sense the temperature drop or the change of lift on their bodies as they leave the miniature thermal, and this triggers their change in direction. There are other paving slabs and hence other thermals, but this is the only one supporting a column of insects. It is also the only one partly shaded by the birch tree under which I sit, and the only one not exposed to a disruptive breeze.

These creatures have found their ideal place and, by the magic of feedback loops, they are staying in it. Incidentally, they are the lucky ones, because every now and then a flurry of wind marks the passage of a full-blown thermal: a large column of rising air, lifting off the nearby fields. If I look up when this happens, more often than not I see a flock of starlings, spinning around in tight turns, taking advantage of the free lift while eating the hordes of unlucky insects that have been caught up in the inverted whirlpool of air. Feedback feeds upon feedback, and every creature seeks out and finds its optimum place in the dance of life.

The flies, even though they behave like whirling dervishes, manage to remain poised in a column of shaded, windless, rising air whose boundary is marked by a tiny temperature change. This is their place. The little green parachutists hatch out and drop from the birch tree to float away on the critically poised, warm, dry summer breeze. This is their day. A few days ago, it was the day of the ants.

The weather then was warm and humid, the air was still and, under the ground, great preparations were taking place. Almost simultaneously, from separate nests as much as 30 metres apart, there erupted volcanoes of flying ants. The ants would emerge from a hole, wriggle their wings free and allow them to dry momentarily in the sunlight. Then they would crawl purposefully to the top of a blade of grass, stretch their wings and fly, apparently using the polarized light from the sun as guidance. They did this in their thousands, and they did it all at the same time. Within each individual nest, I imagine the prevailing weather conditions had triggered the production of some pheromone, or allowed it to build up to an appropriate threshold, and this was the signal that started the aerial exodus.

The remarkable thing, though, was that it happened simultaneously in half a dozen nests, spread out over two separate lawns, divided by a house and various concrete barriers. The message cannot have been carried from nest to nest, so the exodus must have been triggered by the atmospheric conditions alone, which were uniform across the site. There must be a very specific combination of humidity, temperature and wind speed that is ideal for allowing ant colonies to propagate themselves. Evolution has learned this combination and tuned the behaviour of ants to respond to it. The thing that struck me so strongly was how tiny this window of opportunity is. If the suitable conditions spanned a reasonable range, I would have expected some variation in timing across the various colonies. Yet all the colonies made the same decision at the same time. This was their moment, and they were going to take advantage of it.

What poise! What subtlety! Here are larvae that can maintain weightlessness in gently rising air, when the cold English 'breeze' more often than not whips around; flies that can locate and centre themselves above a particular spot when they are obliged to zoom along at a rate of knots, unable to hover; ants that can pick the exact moment when the rapidly changing and highly variable humidity, temperature and wind speed just happen to converge on the critical region in which lies their window of opportunity. It is a harsh world out there, and conditions can be extreme and rapidly changing, yet these tiny loops of persistent phenomena that we call life forms can find their perfect balance. Within the narrow range of conditions that delineates their ecological niche, they can balance themselves with all the poise of a unicyclist. Feedback loops tell them when conditions are right; other feedback loops allow them to remain balanced within the tiniest regions of acceptability. Feedback upon feedback means that one species' poise creates the predictability upon which another species' survival depends (like starlings 'knowing' that food is to be found inside thermals).

A short lesson in humility

I started this chapter with talk of chainsaws and hedge trimmers. My style of gardening generally involves disrupting the balance of nature, rather than contemplating it. I tend to take a 'slash and burn' approach to maintaining the patch of scrubland laughingly known as our back garden. Last week, when the ants decided their moment had come and the lawns erupted in a black, heaving mass, I responded in the way any nature-loving biologist would: I exterminated them. Well, I tried to exterminate them, but clearly I did not succeed. I went around the insect-infested garden, heroically daubing the entrances to their nests with lethal powder. It seemed a fair fight to me: one large and horribly be-weaponed human being, versus several thousand tiny, well-meaning insects.

The poor ants that got hit full in the face by the insecticide obviously expired quite quickly. But the rest simply shut up shop and remained underground. The next day, instead of heaps of dead ants, all I found were patches of unsightly powder. Even those from the initial massacre had disappeared, probably picked off by birds that even now are cursing me and planning vengeance for poisoning their food. The colonies had obviously decided to regroup and wait for a better day. While I was writing the previous paragraphs, I was wondering what had become of them. Had they abandoned their poison-stained entrances and set their workers to dig new tunnels and new emigration points for the remaining winged brethren to emerge from on the next warm, humid day? Well yes, they had. Right under my chair, in fact. Nature has just ganged up and taught me a lesson. You cannot keep a robust persistent phenomenon down. But what timing! Now I'm going to have to move indoors and try to remember what it was I was going to write about. Ouch!

To them that hath is given

So now I am barricaded in the house, while a scene from an Alfred Hitchcock movie is acted out all around me, and I feel secure enough to continue with the topic of this chapter, which is feedback. Feedback is absolutely central to our cause, and it comes in two flavours: positive and negative. In scientific terms nothing pejorative is ever meant by these words. Negative electrical charge is not regarded as inferior to positive charge, and negative numbers are not somehow more depressing than positive ones. However, in everyday life 'positive' and 'negative' do have emotive connotations, and for that reason people frequently misuse or misunderstand what is meant by positive and negative feedback.

People often talk about positive feedback as if it were roughly synonymous with encouragement, while negative feedback is seen as discouragement. Businessmen talk about 'getting some negative feedback from the customers', meaning that the customers did not like the product. Managers might speak of 'giving positive feedback to the staff', implying that the staff should be encouraged and motivated, whether they are doing well or not. In fact, these are incorrect uses of the terms. Positive feedback should actually be defined as 'something that tends to exacerbate a change', and negative feedback as 'something that tends to counteract a change'.

It is true to say that a manager who praises staff for doing good work is providing positive feedback, because the staff are likely to be motivated to do even better as a result – the feedback is thus exacerbating the change. But if the same manager later chastises staff for doing poor work, this does not necessarily turn the process into negative feedback. If chastising them makes them upset and causes their work to decline still further, it is still positive feedback, because the thing that is fed back (the boss's opinion of the staff's performance) tends to amplify the change in both cases. So feedback can be positive, even when what is being fed back is 'negative'.

On the other hand, negative feedback tends to counter any change and restabilize the system. The little green larvae we met at the start of this chapter were using negative feedback to help them stay afloat. If having a short gossamer thread caused them to sink (say) then they lengthened it; if it got too long and they started to rise, they shortened it. In other words, their rate of rise and fall was being fed back in a negative sense to their spinnerets, so that the length of trailing gossamer tended to self-compensate for changes in height.

The two classical examples of negative feedback in action are the thermostat and the steam governor. Thermostats turn heat on when it gets too cold, and off when it gets too hot. They therefore tend to counter external changes in temperature. The governor on a steam engine consists of two heavy balls of metal, swung around in a circle by the motion of the engine. If they swing faster they tend to fly further out, and in so doing they close a steam vent. Less steam is allowed into the chamber and the engine slows down again. Conversely, if the engine slows, the valve will start to open and more pressure will become available, speeding the engine once more. The engine therefore tends to stabilize at a particular speed, despite any changes in the load or the supply of heat.

If the governor were instead designed to make the valve operate the other way round, then a slowing of the steam engine would cause the valve to close even further, and the engine would quickly lose pressure and grind to a halt. On the other hand, if it started off too fast, the valve would open and the engine would accelerate, leading to runaway and probably an explosion. Similarly, if thermostats turned the heat on when it got hot, and off when it got cold, rooms would either boil or freeze. In this respect, simply changing the direction or 'sign' of the information being fed back into the system can change negative feedback into positive feedback (hence the use of 'positive' and 'negative' to describe the processes).

But in practice, negative feedback is not always the simple inverse of positive feedback. It might seem as though the manager described above could convert positive feedback into negative by reversing the response given to the changes in the workers' performance. If the workers get worse, encourage them; if they do well, discourage them, and then perhaps they will tend to stabilize at medium performance. In practice this might work or it might not. If I was underperforming and my boss was delighted with my progress, I would tend to assume that I must be doing better than I had previously thought, so I might well not bother trying so hard in the future. Poor performance plus praise can lead to worse performance. Or perhaps that's just me.

At any rate, things in the real world are rarely simple, and most interesting systems have multiple feedback loops – some positive and some negative, some with a quick response and some that take longer. In the above example, there is a second, slower feedback loop at work called *adaptation*, which complicates things. I am adapting my assessment of how well I am doing because I'm frequently getting rewarded when I'm doing badly. I assume from the mismatch between effort and reward that I must have been underestimating my performance in the past. Consequently I gradually become more immune to encouragement in the future.

The first law of biology

Adaptation is such a common feature of biological systems that I call it the first law of biology: it can be summed up in the phrase 'nature is lazy'. Most biological systems automatically adjust themselves so that they are at rest when the world is in its local average state. In other words, they track a moving average of recent world states and adapt to this so that most of the time they are inactive. They then respond only to deviations from the local average. When astronauts spend long periods in space, their muscles start to waste, because their bodies adapt to the low-gravity environment and do not waste energy on maintaining muscle tissue that is no longer needed.

Adaptation is a feedback mechanism with many uses. The eye, for example, adapts to the average light level it is experiencing, so that brighter than average is perceived as 'bright', and darker than average as 'dark', regardless of the average light level in the scene. This gives it a vastly greater dynamic range than it would have if it tried to respond to absolute light levels. It is such an effective mechanism that when we walk from inside a building to outside, we barely notice the difference and find it hard to believe that the outside might actually be several hundred times brighter.

Neurones adapt too, by reducing their response to a signal if it continues unchanged for a long period. This way they don't use up energy when there is nothing to report, and they remain able to respond to tiny changes, regardless of whether that change is from zero or from an already high signal level. Even conversation levels in a bar will adapt to the local average, but since everyone is trying to be heard over the top of this average, bar-room noise levels often climb incessantly upwards. I remember a particular San Francisco restaurant in which we had to put our mouths close to each other's ears and scream at the tops of our voices to make ourselves heard – an arms race between conversations caused simply by the ambient music, which had been set at a level higher than that of normal chatter. Once you learn to recognize it, you can see adaptation everywhere.

So adaptation is one of the key 'clever tricks' by which nature exploits feedback. Later, we shall use adaptation to help us design a creature that can learn. In the example about the manager trying to control the staff's performance, there were two loops: negative feedback on a long timescale (adaptation), combined with positive or negative feedback on a short timescale (depending on whether the manager chose to amplify or diminish the changes in the workers' performance). Predicting the overall behaviour of these nested, intertwined feedback loops is not at all straightforward, and this is what makes the natural world so interesting. But if we are to play at being Frankenstein, and want to make complex adaptive systems that are good enough to be called alive, intelligent or conscious, then we need to understand complex networks of feedback loops and learn how to manipulate them.

Feedback has its ups and downs

In fact I can think of no more important a topic of enquiry. After all, understanding the behaviour of complex loops of cause and effect is what intelligence is for, and our ability as human beings to perform this trick and predict the future from our knowledge of causal relationships is what enables us to survive. The formal study of feedback is known as cybernetics, and one might think that it was so important that it would be taught in schools. After all, the same basic feedback processes underlie so many things – economics, biology, electronics, engineering, sociology, geology... Learning about the principles of feedback systems could help children to understand so much at a single stroke, and provide coherence to otherwise disparate subjects. Sadly, not only is cybernetics not on school curricula, it is barely studied anywhere any more. Only a very small number of cybernetics departments exist in universities, and they tend to focus rather heavily on electrical engineering or economics.

Fifty years after its birth, cybernetics has mysteriously almost faded from view. In its place, though not really a replacement for it, stand the closely related fields of complexity theory and dynamical systems theory. These more modern disciplines are really what one might call meta-theories. Instead of dealing with what feedback-driven systems *actually* do, they are more concerned with what kind of things they *tend* to do. In other words, they are concerned with the probabilistic behaviour of complex systems, not with the details. Complexity theory is thus to cybernetics roughly what traffic analysis is to automobile engineering. Each has its place, and I think it is a real shame that cybernetics has managed to get itself into such a rut, just when society has real need of it. Happily, one thing that complexity theory is really good at is helping us to visualize things. It is especially rich in metaphors based on geography, and one such set of geographical tricks can help us to visualize the key characteristics of feedback systems.

Although it is not an altogether reliable analogy, we can visualize feedback loops as a *landscape* of hills and valleys. To begin with, imagine a cross-section through a landscape, as shown in Figure 12. Here, negative feedback would be represented by a valley or hollow. If the current *state* of the system is visualized as a ball, free to roll around the landscape, then it is obvious that gravity and the hollow surface

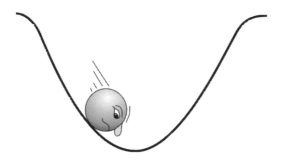

Figure 12. Negative feedback.

combine to create negative feedback. If something perturbs the system by moving the ball sideways, it will find itself forced slightly uphill, and gravity will cause the ball to roll back in the opposite direction, towards the bottom of the hollow again. It may overshoot a little, and suffer a few swings from side to side before finally coming to rest at the bottom. This overshooting is known as *ringing*, and is characteristic of negative feedback systems if they are not *damped*. A damped negative feedback loop would be like a hollow whose sides are made of sticky clay. In this case, the ball would have less of a tendency to overshoot, and would come to rest quickly and with less ringing. Notice also that there is a special spot in the landscape – the place where the ball tends to come to rest. I call this the *null point* of the system. When the ball is to the left of the null point, it has a tendency to swing right; when it is to the right, it swings left. All feedback loops have a null point.

If a negative feedback loop looks like a valley, then you might expect a positive feedback loop to look like a hill, and you would be right. Positive feedback has a null point, too. In this case, it is the very crest of the hill. Place a ball exactly at this point (i.e. put the system into this state) and nothing will happen. The ball has no net force acting in either direction. However, perturb the ball only slightly to one side, and it will start to roll down the slope. As it falls it will gather speed and fall even faster. This is positive feedback in action. Once the ball has made a choice to move in one direction, no other options are open to it, and it will accelerate rapidly; sooner or later, it will hit rock bottom and stay there. Positive feedback loops therefore have *end points* (or end states), as well as a null point. A system with positive feedback will tend to *lock in* to one or other end state. If it manages to remain poised on the null point then it is a fluke, because even the slightest disturbance will cause the system to lock into an end state.

We can visualize systems of combined feedback loops as a complex landscape of hills and valleys. If our ball rolls down from a positive feedback hill, it may come to a complete halt in a wedge at the bottom of a slope, or it may fall into a rolling valley of negative feedback, and ring for a while, before coming to a stop. Perhaps the hill on the far side of the valley is lower than the one where our ball started, and the ball will have enough energy to continue up the other side and over a second null point into another phase of positive feedback.

One very common scenario in nature is the valley-on-a-hill (Figure 13), created when a positive and a negative feedback loop coincide. In such a situation the ball would stabilize in the middle of the valley after any perturbation, but only as long as the perturbing energy isn't large enough to carry the ball over the lip of the valley and off into the fatal positive feedback slope beyond. Once this happens, the system swings violently into an end state from which recovery is impossible. An example of this valley-on-a-hill combination is the so-called poverty trap. If I have less than a certain amount of money, I shall not be able to make ends meet and I shall get into debt. Paying the interest on this debt will then make me even poorer, and I shall have to take out further loans to pay off the first one. On the other hand, if I have more than a certain amount of money I can afford to invest some, and so will gain even more money in the future. The threshold between these two scenarios is the null point of a positive feedback loop. At the same time, for most of us our income roughly matches our outgoings and we have little flexibility available to us. If we have overspent one month we can

Figure 13. Valley on a hill.

Figure 14. A feedback landscape.

be more careful the next. Alternatively, if we have a little spare cash one month we are unlikely to invest such a small amount, but will treat ourselves to a little spending spree in the following period. This self-compensation is a negative feedback loop, and most people manage to remain in the safety of its walls – never getting poor enough to slide off into debt, but never getting rich enough for their wealth to snowball either.

Taking such landscape analogies too far can be dangerous, but they can certainly help us to picture otherwise difficult mathematical relationships. Stretching the analogy a bit further, it should be obvious that our two-dimensional landscape is acting like a graph, where the horizontal axis represents the possible states of the system, and the hills and valleys on the vertical axis tell us into which states the system is likely to fall and which ones it will tend to avoid. Most real systems have more than one variable – more than one thing that feedback can alter – so representing these systems requires more axes. A system with two variables could be represented as a surface, looking rather like a relief model of a landscape (Figure 14), rather than a simple cross-section. Now the ball has more interesting properties. As well as rolling left and right, it can roll into and out of the scene. Changes in one axis (variable) can result in changes in the other. The repertoire of hills and hollows is now enriched by other topographies, such as valleys, corries, ridges and saddles. What we are looking at here is approximately what is known as the *phase landscape* of the system. In the jargon of complexity theory, the end points of positive feedback and the null points of negative feedback, as represented by the hollows and crevices on our surface, are called the *attractors* of the system, because the system tends to fall into these states. A huge amount of metaphorical and mathematical value can come out of this topological approach to studying complex systems, and as a non-mathematician I think it's one of the most delightful mathematical models ever conceived.

If you want to represent a system with more than two variables (and remember that a living organism might represent a system with zillions of them), you have to move into hyperspace. We have taken up all three spatial dimensions to represent a system with two variables – the third dimension is the height of the surface and describes the way that those two variables interact. For three variables we need three axes, plus a fourth to describe the relationships, so the variables could perhaps be represented as a cubic volume, and the dynamical behaviour of the system by changes in density within this volume. Unfortunately the analogy of the ball fails us miserably at this point. With four variables we are into four-dimensional hypercubes, and my imagination abandons me completely. Happily, for most purposes, sticking to a surface of hills and valleys is enough – we cannot represent reality on such a limited canvas, but we can understand all the important principles.

It doesn't rain all the time, even in England

One last demand can be made on our landscape metaphor. Feedback loops can coexist in different regions of the landscape, as I have already described – the ball can roll off one positive feedback hill and ring for a while in the valley at the bottom of the slope, where a negative feedback loop has taken over. Feedback loops can also interact in more subtle ways. It is quite possible to imagine different *orders* of feedback, where the feedback function of one loop is *modulated* by that of another. For example the *gain* of a negative feedback loop (the steepness of the slope, and therefore the rate at which the system collapses back to stasis after a disturbance) could be modulated by another feedback loop, responding to some other variable.

This sort of thing happens in weather systems, for example. Clouds are built from positive feedback loops, as we have seen. Once they start, they tend to accelerate and the clouds get thicker and thicker. But we are not permanently cloaked in cloud, despite the apparent tendency of the system to lock into a state of total cloud cover. This is because there is also a negative feedback loop at work, controlling the gain (the level of amplification) of the system. As the amount of cloud builds up, the amount of sunlight hitting the ground is decreased. This reduction in light lowers the strength of the positive feedback loop, and thus slows down the rate of cloud production when it threatens to take over. The system tends to dance around a happy medium, sometimes with more cloud and sometimes with less, but rarely with total cover and rarely with none at all. If there were only a single, positive feedback loop, then the result would be total cloudiness for ever. With an additional negative loop, the system settles out at a happy medium. That this happy medium can also vary from day to day is evidence for a third loop, caused by the changing amount of water vapour, which is itself controlled by the amount of sunlight hitting the ground over a larger area. Here we have the balance of nature in all its glory: of all the possible moisture levels our atmosphere could find itself in, the region in which both clouds and blue sky can exist simultaneously is very narrow indeed, yet the skies are constantly kept in this region. The sky is an ever-changing, restless adaptive system which not only keeps our planet from freezing or boiling, but also provides the greatest free show on Earth.

So loops can interact on different timescales and at different levels or orders. On our landscape, the easiest way to visualize this is to add time to the equation, and imagine that our hills and valleys are constantly changing shape. It is almost as if the position of the ball (the state of the system) is deforming its own landscape, and thus its future attractors. In fact, this analogy works best when we imagine several balls on the one landscape. Several living things, for example, can occupy the same environment or share the same physiology, and therefore can be described by the same landscape. Each individual will be in a different state at the same moment, and so each can be represented by a different ball. If two balls somehow communicate, or otherwise interfere with each other, then it is possible to imagine that one disturbs the landscape on which the other is rolling. This makes the most sense when we consider 'evolutionary landscapes', where the hills mean something else entirely (fitness levels, not feedback loops). Nevertheless, it can be helpful in a feedback landscape too.

Running the ridge

The story of an organism's life can be seen as a journey across such a landscape of positive and negative feedback paths. Imagine a real land-scape – a sort of badlands – in which wide plains have been cut from the hills, leaving a series of interconnecting ridges. Imagine yourself as an organism, and start out at the top of one of these ridges. To either side of you lies death. If you move too far to the left or right, you will start to slip down the slope into the valley below. If you wander just a little off course, you may have sufficient energy resources to turn around and climb back up the hill, but once you have gone too far and started to slip too quickly, there will be no way back.

If that were not bad enough, now imagine that the part of the ridge you are standing on is subsiding. The longer you stay still, the more untenable your situation becomes. You have no choice but to run forward for ever. You dare not stay still, but you cannot see more than a few steps in front of you. Which way should you go? If you feel your way forward, away from the crumbling ground under your feet, you might perhaps begin to tell whether your path is taking you away from the ridge. If you turn to the right, perhaps you will start to descend even faster. So you turn left instead, and hopefully you have sufficient energy to regain the ridge. If you turn too sharply you will overshoot and start to descend the other slope, this time with your own momentum sending you even faster towards oblivion.

Even if you judge it just right, and can zero in on the direction of the ridge, you have to contend with the fact that the ridge itself is not straight, and the path that was once correct becomes treacherous as the crest of the ridge dives off to one side. Worse still, you are not the only organism running the ridge – others are too, and as they slip and slide they cause small landslips and deform the landscape in front of your feet. If you are smarter than the average organism, you will not look down at your feet and hope to react in time to your mistakes in direction. Instead, you will look ahead. You will try to see and remember how the ridge twists and turns in front of you, and you will try to adjust your balance to prepare yourself for what is to come. Perhaps you will even allow yourself to dive off down a slope, knowing that the ridge turns around a little farther on and your trajectory will carry you on and up the next slope, back onto safe ground. That way you can make best use of your fading resources, by using the slope to slingshot you in the same way that an interplanetary spacecraft swings itself around Jupiter's gravity well to gain energy.

Looking ahead at the slope is one thing, but remember that the other organisms are changing the landscape in front of you. If you are really smart, you will plan your movements to take account not only of the curving ridge, but also of how you believe the other organisms will react, and therefore what damage they might do to the path.

What you have just experienced in your imagination is what it is like to be alive. We all live life on the edge: we have limited energy resources available to us, and death awaits us round every corner. The challenge is to keep on the ridge, and avoid slipping so far that we can't recover. The ability to control the flow of energy and change course in response to information about whether we are sliding downhill is what is called adaptive behaviour. All living creatures can do this, but only some can take the next step, which is to look ahead. Looking ahead enables an organism to predict the need for future changes of direction and plan for them. This ability to learn the relationships between cause and effect and to use them to predict the future is what is called intelligence, and the best and brightest intelligent systems can make plans that are more than just 'reactions in advance'. These systems can reasonably be called creative, if not conscious. They can make 'what if …' judgements that lead to innovative ideas like risking death to take advantage of the slingshot effect. Finally, there is the ability to imagine what the effect of other organisms will be on the environment in front of you. At a minimum this requires that you can predict highly non-linear sequences (not just where the slopes are now, but where they will be when someone else has run over them). At a higher level still, it requires the ability to put yourself in someone else's shoes – 'If I were them, I wouldn't try to stay on the ridge there. I'd slingshot myself round that corner, and if they do that, I predict that the landscape will deform because of their action; therefore I had better go this way.' This ability is what psychologists call theory of mind.

Notice that each solution to the problem requires feedback. The simplest form (adaptive behaviour) requires relatively straightforward negative feedback, to compensate for mistakes in direction. Without damping, though, the organism may overshoot, and since the ruggedness of the terrain is unknown and ever-changing, even the level of damping may need to be controlled by feedback. Intelligence thus requires several orders of feedback. Some of them control behaviour, while others control *changes in behaviour*, a process we call learning. More advanced intelligence requires feedback from events that haven't even happened yet. This is quite a feat, and something we shall look at more later on.

The outside world can be visualized as a set of feedback loops as I have just described, and I find it very interesting that the inside world can too. In this 'inner space', feedback mechanisms not only enable a creature to learn, but also represent the actual result of that learning – these are the loops that determine the creature's responses to its environment. In an important sense these internal feedback loops are a mirror image of those outside, because their job is to compensate for the external forces that are threatening the organism's ability to persist. If a positive feedback loop exists in nature, such as the poverty trap, then the brain's correct response to it can be represented by an equal and opposite negative feedback loop (a rule such as 'cut back on expenditure if money is running short'). Because the internal feedback loops are a reflection of the world outside I think it is fair to call them mental models of the world. My present research is focused on working out how general feedback-supporting devices such as neurones can develop this ability to build usable models of an organism's external feedback landscape.

The first time this landscape metaphor for life really struck me was when I was hill-walking in the Derbyshire Pennines. Next time you find yourself up on the hills among ridges and saddles, you might want to do what I did: wait until no one is looking and then, safety permitting, try hurtling along as fast as you can, without getting so far off the ridge that you start to slide into the valley. Try it while looking down at your feet and notice how slowly you have to go. Then try looking ahead, and perhaps try planned excursions down the slopes to slingshot yourself at minimum effort up the next part of the ridge. You may not learn much about life, but you'll certainly develop a healthy respect for it!

Disinformation

When we speak of feed-forward and feedback, what exactly *is* it that is being fed forward or back? Any good technical answer to that question is likely to contain some or all of the words, 'signal', 'transmit' and 'information', as in 'a signal is being transmitted around a circuit, causing information to flow from place to place'. This is a good answer but, as always, these words come with some baggage that we need to be wary of.

The transmission of a signal, for example, should not be confused with the movement of material. When we think of electrical signals being passed down a wire, it is easy to assume that the signal is the same thing as the current – that the electrons are carrying the signal from place to place as they squirt down the wire. However, electrons move relatively slowly through metal, while electrical signals travel at near the speed of light. It can be helpful to visualize electrons in a wire as a necklace of metal balls connected by springs. When the ball at one end of the chain is disturbed, it causes a wave of disturbance to travel down the necklace, even though the necklace itself is not necessarily moving along at all. Another analogy is a domino run – knocking over the first domino causes a wave of activity to pass down the entire length of the chain (if you're lucky), even though no individual domino travels more than few centimetres. Signals are therefore not stuff. They are non-physical, persistent patterns with a coherence and existence of their own. Like so many other entities in this book, signals are non-things.

Information is a rather trickier concept with its own pitfalls. In 1948, Claude Shannon and Warren Weaver from Bell Laboratories published a paper that defined a new science called information theory. This theory was a godsend for telecommunications engineers, and has turned out to have profound implications for many other fields too. Some of the technical terms have even found their way into common usage, as in 'I'm sorry, I can't deal with that now – I just don't have the bandwidth.'

In Shannon's theory, all information can be measured in bits – the number of binary digits needed to represent the information content of a signal. The information content of a traffic light, for example, requires no more than three bits. Each bit could represent the state of one of the three lights: a bit would be 1 if the corresponding light was on and 0 if not. Red would thus be 100, amber 010 and green 001. Red and amber simultaneously would be 110. In fact, three bits is more than we really need, because in practice traffic lights don't make use of all eight possible combinations of colours. A British traffic signal uses four of the possible combinations (most other countries use only three), so two bits are sufficient to represent each of the meaningful states as a binary number from 00 to 11 (0 to 3 in decimal). Here we meet an idea that is crucial in information theory: the information content of a signal can be measured in terms of its compressibility. Although our traffic lights could be logically and conveniently coded using three bits, two will actually suffice. The data could thus be compressed from three to two bits, requiring 33 per cent less bandwidth (measured in bits per second) to transmit the information as a signal.

When we work with visual images in a computer (a photograph on a Web page, say), those images are usually stored in a compressed form. All the superfluous, redundant data have been removed, and the remainder have been encoded in a minimal or near-minimal way to reduce the size of the image file. Since compressibility is considered to be a measure of the information content of a signal (or an image), a photograph that can be compressed to one-tenth of its original size is regarded as having less information in it than one that started out the same size but could be compressed by only 50 per cent. This seems a very rational and useful idea. It is easy to imagine that a photograph of a crowd scene is both less compressible and more information-rich than a photograph of a cube. The concept remains useful, even though it leads to the rather counter-intuitive conclusion that the most information-rich signal of all is therefore random noise (because noise has none of the regularity or predictability that would enable it to be compressed).

Nevertheless, this conception of 'information' can be misleading or inadequate if taken on its own. Look carefully at Figure 15. This shows two pieces of ultra-pure silicon that have been 'doped' with tiny amounts of impurities, in each case arranged in very slightly different

Figure 15. Which is the overcooked sand?

patterns. What is the significant difference between the two objects? Well, according to information theory, there is no difference at all. The pattern on each silicon slab turns out to be equally compressible, and therefore each must contain the same amount of information. Yet it seems to me that there is a crucial difference, which information theory does not account for. Although the left-hand object is nothing more than a slab of dirty silicon, the one on the right is a computer. Information theory tells us that the two are equivalent in their information content, and yet a computer chip is clearly very much *more* than a piece of dirty silicon. They may contain identical *amounts* of information, but there is something distinctly special about the information content of the chip on the right.

Instead of two silicon chips, I could have shown you a statue and a lump of clay, and arranged for the descriptions of their structure to have the same level of compressibility and therefore the same information content. If I had, you would probably have felt that the statue still 'had something' that the lump of clay did not, and you might have decided that this extra something was 'meaning'. Lumps of clay are meaningless, but statues mean something to people. Since a statue means something to a human being and maybe to a sheep, but clearly means nothing at all to an ashtray, we seem to have something here that a physicist would call an observer effect. It is as if a statue contains something special *because* a human observer can ascribe meaning, or perhaps *utility* to it. This may be true of statues, but it does not seem entirely satisfactory to me as a general explanation (observer effects are troubling because they often imply vitalism). A computer chip surely possesses something in an absolute sense that a slice of dirty silicon does not, regardless of whether a human being is there to witness it.

Whether this utility is absolute or whether it is meaningful only in the eyes of a human observer, we have here an appealing concept that I can only describe as 'elegance'. A thing is elegant if it maximizes some measure of utility while minimizing the information content (i.e. maximizing its compressibility). Elegance is therefore proportional to utility multiplied by compressibility. Another way of looking at this idea is to distinguish between complexity and complication. Something is complex if it contains a great deal of information that has high utility, while something that contains a lot of useless or meaningless information is simply complicated.

Utility may also be an observer effect, but in the example of the computer we can see something that is potentially absolute and measurable. The computer chip may contain only as much information as the slice of dirty silicon, but computers can *generate* information, and the potential information embodied in a computer is huge compared with that in a lump of overcooked sand. Perhaps even statues have information potential, since they can evoke responses from human beings and so change the universe in ways that heaps of clay fail to. Certainly 'information' in the non-technical sense of the term has a qualitative dimension, as well as the quantitative one that information theory gives it.

Finally, even the word 'transmitted' has its problems. Transmitting implies an active role and receiving implies a passive one, and that can lead to fallacies about the concept of control. As I have said before, control is not synonymous with domination. It is just something that happens – an effect as much as a cause. Thinking of control in systems as being some kind of manipulation from on high is a dangerous and misleading idea, but it is one that our hierarchical, pugnacious society beats into us from an early age. A feedback loop is just that: a loop, with no end and no beginning. Unlike people, things do not tell one another what to do. Some shout and others listen, but that is not the same thing. In fact, one of the truly important tricks that computer science can learn from nature is to separate the shouting from the listening, so that the shouter need not know about the listener and vice versa. My adrenal glands do not tell my heart to beat faster. Adrenal glands respond to signals by secreting adrenaline; the heart responds to adrenaline by beating faster. Adrenal glands do not know what hearts are, and hearts have never heard of the endocrine system. Adrenaline hasn't heard about either.

This is the way organisms work. There is no architect, and no master controller telling the system what to do. There are just vast numbers of small independent entities that respond to signals as and when it suits them, and emit new signals whose destination they do not know. Top-down control leads to complexity explosions, because something somewhere has to be in charge of the whole system, and how much this master controller needs to know increases exponentially with the number of components in the system. Living systems are bottom-up: no part knows or cares what its role is in the whole, but the whole still emerges from the cacophony of these zillions of mindless loops of cause and effect.

I hope I have given you some ways to visualize complex systems of interacting feedback loops, and at least a flavour of the perpetual domino run that is the universe. Feedback comes in only two varieties, but by varying the null points and timescales of feedback loops, and using one feedback loop to modulate another, systems of enormous complexity, subtlety and persistence can be realized. You and I are living proof of that. But there are still things to be said about exactly how one flow of causality can affect another, what fundamental 'techniques' (such as adaptation) nature uses, and what for. We need to create a Lego set of parts that can be plugged together to make something that lives, thinks and feels.

Now that the stage is set, it is time to introduce the cast.

CHAPTER NINE

* *

GOD'S LEGO SET

In this layer the cells are polygonal, triangular, or fusiform in shape.
Each polygonal cell gives off some four or five dendrites, while its
axon may arise directly from the cell or one of its dendrites. The axons
and dendrites of these cells ramify in the molecular layer.

Henry Gray, FRS, *Anatomy, Descriptive and Surgical*

Just as a thermostat keeps the temperature of a room constant in the
face of external perturbations, so life uses feedback to maintain itself,
to adapt to change, to grow and to reproduce its pattern indefinitely
across generations. But although some living creatures may seem
barely more sophisticated than thermostats, the task facing them is in
reality nowhere near as straightforward.

Take *Domesticus suburbii,* for example. Commonly known as the couch
potato, this primitive species maintains a homeostatic (self-regulating)
balance by the simplest possible means. When it finds itself getting bored,
it just reaches out and changes the TV channel. When it is hungry, it
simply increases its intake of popcorn to compensate. Yet even these
straightforward reflexes are far more sophisticated than they might
appear at first glance. The function of a thermostat is to send out a correction signal based on the difference between the current state and the
desired state. If the room is colder than it should be, then the thermostat
switches on the heater; if it is too hot, the heater is switched off. Although
the couch potato achieves a similar effect by minimizing the difference
between its current and desired states of hunger or boredom, it cannot do
so by sending out a simple error correction signal. Knowing how far it is
from the ideal level of hunger does not in itself tell *D. suburbii* how to eat.
An extreme variety of couch potato might connect itself up to a drip feed,
and so be able to regulate its beer intake in proportion to its degree of

thirst merely by opening or closing a valve. But even turning a valve requires rather complex muscle movements. More sophisticated homeostatic behaviours, such as phoning out for pizza, require incredibly long sequences of carefully scheduled muscle contractions, many of which themselves involve feedback control. And yet the couch potato is only one of many lowlier kinds of life form. More advanced species, such as frogs or spiders, may have to carry out far more elaborate foraging, mating or defensive activities in order to survive.

Living things, then, are homeostatic systems like thermostats or steam governors, but differ from these mechanical devices in that the actions they need to take to correct any imbalance are not usually simple or obvious, and may have to be learned through experience. The net result of an organism's behaviour is negative feedback, through which it maintains the status quo in the face of a variety of environmental stresses. But the implementation of that overall feedback path requires the interaction of many intermediate feedback loops, some negative and some positive. How can such simple flows of cause and effect combine to produce the intricate sequences of activity that allow life to maintain itself?

The quote at the head of this chapter is from *Gray's Anatomy*, which I like to think of as the bible of old-fashioned materialist biology. Anatomy divides organisms into identifiable building blocks at the material level, from organs to tissues, from types of joint to bulges in the physical appearance of the brain. But there is another way to divide things up, and a different set of building blocks that we might identify, based not on physical structures but on relationships between cause and effect. If we think of an organism as a network of feedback loops instead of a physical collection of cells and tissues, then a rather different pattern of common features emerges. In this chapter I shall identify some of these components, with the aim of defining a general toolkit for building living things. Later, we shall attempt to put these bricks together to create life.

One small boy and a herd of rampaging bulls

We have seen that a feedback loop is a flow of cause and effect that modulates itself. In a negative loop, some of the signal is fed back in a way that counteracts change. In a positive loop, on the other hand, it is sent back in a way that reinforces the change. But what kinds of mechanism will allow one signal to modulate another? If you have ever worked with electronic components, you will know one of the answers to that question: the transistor. Transistors and thermionic valves (vacuum tubes, if you're American) are both devices that allow one electrical signal to modulate another. In a valve, the flow of electrons across a vacuum is modulated by having the electrons pass through a grid of wires on their way from one metal plate to another. When a voltage is applied to the grid, the grid fills up with other electrons which repel the ones trying to flow through it, like guards defending a castle against invaders, and so attenuate the flow of current. A transistor works in a similar way, but the action takes place in solid silicon that has been modified to increase or decrease the number of free electrons in different regions.

In the transistor, a tiny change in the voltage on the base (the transistor's equivalent to the valve's grid) can make a much larger difference to the current flowing through it. It is this amplifying effect of transistors that is used in stereos and public address systems. When I was young I used to imagine a transistor as being like a small boy sitting astride a fence and controlling a sliding gate between two fields. A herd of rampaging bulls might be rushing through the gate, but the small boy could easily reduce or increase the flow of these huge animals by sliding the gate to and fro (in fact the controlling element on a certain kind of transistor is called the 'gate', so my imagination was not too far off the mark).

A transistor can use one flow of cause and effect – a signal bounced from electron to electron along a wire – to modulate another flow. This becomes feedback when the flow that is doing the modulating is also the flow that is being modulated. If some of the output signal is fed back to the base, then the flow controls itself. But transistors are not the only things that do this. Customs duties modulate the flow of goods into a country, traffic lights modulate traffic, enzymes modulate the production of chemicals. The same basic principle turns up in many situations under different names. Each example uses a different mechanism, but the effect is the same. It deserves a name of its own – let us call it a *modulator*. It is our first and simplest cybernetic Lego* brick.

* Lego is a registered trademark of the LEGO Group.

Broadcasting to the nation

Some flows of cause and effect are very localized, while others are widespread, so modulators are found in both directed and diffuse forms. A transistor controls the current flowing along a narrow wire, so it is a directed form of modulation. An enzyme let loose in a sea of chemicals modulates the concentrations of the chemicals whose reactions it catalyses. Each individual molecule and catalytic event is spatially discrete, but taken en masse the modulation is widespread.

Directed and diffuse information flows have different characteristics, and feedback systems (especially living ones) often use both. They are equivalent to the telephone and the radio. Many telephone conversations can take place simultaneously, and each can be routed to a specific destination without the whole thing turning into a cacophony. Radio signals are not so selective, but they can reach many places at once – they provide global communication and allow one thing to speak to many. In biological systems, directed communication is predominantly by electrochemical signals sent along nerve fibres (although the action of a muscle on a joint can equally be thought of as directed modulation). Diffuse signalling is usually performed by chemicals that are spread through the bloodstream, through the tissue fluid between cells or among neurones in the brain. As well as differing degrees of selectivity, the two forms tend to have other unique characteristics: nerve impulses are rapid, while chemical diffusion takes time. Diffusion also generates *gradients*, where the concentration of a chemical signal varies smoothly from place to place – a kind of halfway house between global and local information flow.

Both directed and diffuse forms of communication are important, yet scientists who study the brain, especially those who try to emulate it in artificial systems, sometimes ignore diffuse communication altogether. This is despite the fact that our heads are flooded with neurotransmitters and neuromodulators that sometimes travel quite large distances inside the brain. I suspect that this is a really big handicap to those trying to understand how the brain works, and we shall certainly use both forms when we come to make our own little brains.

The brain is commonly thought of as the only 'computational' device in the body, but the body's chemical system is also a computer of a kind. The chemical transformations that take place during digestion, pregnancy or an immune response are also information transformations, and a chemical reaction is similar to a mathematical function. This is because a reaction produces an output (the rate of production of a chemical) which is some function of the concentrations of its input chemicals. As far as the body is concerned, networks of chemical reactions are mostly ways of converting raw materials into useful substances, but the brain can also use the information provided by such reactions to establish useful facts. For example, our brains need to be aware of how hungry we are, and this information is not directly obtainable from the environment. But it can be determined by monitoring the levels of intermediate products from the chemistry of the digestive and respiratory processes. In a way, thinking extends beyond the brain and makes use of diffuse computational transformations in other parts of the body.

But back to our Lego bricks. For an organism to use a combination of directed and diffuse signalling it must have some means of converting from one to the other. Our second class of building block is the *transducer* – a generic term for a mechanism that changes one kind of information into another. Microphones, loudspeakers, car indicators and noses are all transducers.

Sometimes this transformation is directed-to-directed, or sometimes the conversion is between directed and diffuse. A neurone in the brain that emits a neurotransmitter is carrying out a transduction between an electrical signal and a chemical one. Many of these neurotransmitter chemicals travel only a tiny distance across the synaptic gap between two neurones, before being transduced back into electrical form. Sometimes, though, these neurotransmitters diffuse out from their source and have more distant effects. This is a directed-to-diffuse transduction. Similarly, when a diffuse-acting brain chemical (like nitric oxide, or heroin, or any of the hundreds of neuroactive substances) triggers an impulse from a neurone, this is a conversion from a diffuse to a directed form. Such changes in selectivity do not happen only in the brain. When a sound is converted to nerve signals by the ear, this is diffuse-to-directed; when a nerve signal to the pituitary gland causes it to secrete a hormone into the bloodstream, the transformation is directed-to-diffuse.

When we design a life form of our own in the coming chapters we shall be using two specific types of signal: nerve impulses will be our main directed form of signalling, and chemistry will be the main diffuse form. So for our purposes two principal kinds of transducer are needed. These versatile building blocks I shall call *chemoemitters* and *chemoreceptors*. Natural organisms have many different mechanisms for performing these tasks, but we can manage with only two general-purpose Lego bricks. The general-purpose chemoemitter produces a specific chemical at a rate proportional to the nerve impulses that feed it. The chemoreceptor, on the other hand, produces nerve impulses at a rate determined by the concentration of a given chemical. By attaching these two types of building block to different structures we can convert between nerve signals and chemistry and produce a wide variety of useful effects.

You can really feel the difference

Brick number three is the *differentiator*. Sometimes it is not the absolute level of a signal that matters, but the rate at which it is changing, or even the rate at which the rate of change is changing. Knowing the position of a charging rhino is not as important as knowing its speed and direction. Executing a tennis serve depends not so much on the present speed of the falling ball, but the rate at which it is accelerating, and therefore on the time at which the racket should be in a position to intercept it. Differentiators are found in many places in the natural and artificial worlds. They all function by making some form of comparison between the present level of a signal and its previous level. This implies some kind of storage mechanism, or memory of the past, to provide the benchmark against which to make the comparison.

One of the reasons why a differentiator might be useful to a biological system is to measure whether a signal is increasing or decreasing. For example, it is possible for a pair of chemical reactions to produce some substance whose concentration is a measure of how quickly the concentration of another substance is rising. This chemical might then trigger a chemoreceptor in an organism and initiate a response from it. Such a circuit might evolve to act as an early warning system, allowing the organism to take preventive action before the situation becomes untenable.

Instead of comparing the present value to that which immediately preceded it, many differentiators use a reference signal that adapts slowly over time, allowing them to compare the signal at any one moment with a moving average of all recent signals. An organism can thus tell whether something has suddenly changed. You can see this happening before your very eyes by staring fixedly, without blinking, at an object in front of you. If you can resist your eyes' natural tendency to flick from place to place for a minute or more, you will find that your whole visual image starts to turn grey. This is because the cells in your retina and the visual system of your brain are adapting to the average signal reaching them. When you are staring at a constant pattern of light and dark, the average of all recent signals soon matches the instantaneous signal exactly, so the cells stop reporting anything. The moment your eyes move even slightly, the new signal reaching each part of the retina suddenly moves away from the baseline average, and it is this difference that is sent to the brain, causing the image to jump out at you quite dramatically.

Little and often

The complement of differentiation is integration. Instead of detecting the changes in a signal, an integrator measures the total *amount* of signal arriving over a period of time.* Some signals provide information in two ways simultaneously, by varying in both size and rate. For example, in a noisy environment we find it more unpleasant as the noise gets louder, but also as it gets more frequent – the total level of discomfort is (within limits) proportional to the total amount of noise energy reaching our ears over a period, not just its loudness. Integrators can capture this total value by summing all the individual bursts of signal received over a period.

Sometimes integration is used to convert a frequency-modulated signal into an amplitude-modulated one. Nerve impulses are sent as streams of short 'spikes', and the closer these spikes are together, the more imperative the signal. When nerve spikes reach a muscle, on the other hand, what was previously encoded as a spike rate now has to be

---

* Strictly speaking, this is a description of a 'leaky integrator', because a proper integrator would measure the total amount of signal over an infinite period, whereas a leaky integrator has a shorter memory.

converted into a steady force whose strength is proportional to the spike frequency. Because muscles take some time to relax after a nerve spike has caused them to flinch, the closer the spikes are together, the less time the muscle has to relax. Muscles automatically act as integrators, therefore, turning a spike rate into an intensity.

The easiest way to visualize an integrator is as a storage vessel with a leak in it. Imagine a leaky bucket into which you pour random amounts of water at random intervals. If the total volume of water you add to the bucket over a period is greater than the rate at which water leaks out through the hole, the bucket will steadily fill up. This is almost an integrator – but not quite. What we need to do is ensure that the leak is through a vertical split in the side of the bucket, rather than a hole in the bottom. Then, as the bucket fills, the area of the leak under water increases, and the leakage rate goes up. Sooner or later the rate of leakage will match the average rate of incoming water, and the level in the bucket will settle at a fairly constant height. The level of water in the bucket is now a measure of both the size of each splash of water entering it and of the rate at which the splashes arrive.

Integrators are one of the most useful biological building blocks, and they turn up all over the place. Neurones, for example, make great leaky integrators. An incoming train of nerve spikes serves to 'charge up' the neurone, which is simultaneously leaking charge away. If the level of charge settles down to something higher than the neurone's threshold, it will fire and produce an output spike (or spikes) of its own. This burst of output energy reduces the neurone's charge to below the threshold value, and the process starts again.

Chemistry is also good at integrating information. As we saw with our shoe-swapping analogy in Chapter 4, the rate at which a chemical reaction proceeds depends on the concentrations of the reactants. So if a 'signal' chemical is converted into a 'storage' chemical by one reaction, then 'leaked away' again by a second reaction, the concentration of the storage chemical will settle to a level that represents the average concentration of the signal chemical over a period of time.

Leaky integrators are building blocks that provide us (and nature) with a number of useful computational features. As well as integrating signals over time, they act as a memory (because the internal state is a reflection of past inputs) and they provide useful damping. This is because the faster you pour the signal in, the faster it gushes out. So the more charged-up the integrator becomes, the harder it is to charge it any further. In a system with many sources of amplification, this 'back-pressure' is important to prevent the whole thing from locking up.

Incidentally, integrators are frequently used in electronics for all the above reasons, and they are easy to make. The most common use for such a circuit is as a low-pass filter: slowly varying signals can get through easily, while higher frequency ones get 'soaked up' by the integrator. The bass control in your stereo is an integrator which allows low-frequency bass signals to get through more effectively than those with higher pitch. Similarly, the treble control is an electronic differentiator circuit that acts as a high-pass filter. Because differentiators are sensitive to rates of change, they allow rapidly varying signals to pass through more readily than slow ones.

See-saw

You might imagine that any randomly wired network of the building blocks described so far would result in a system that settled down into a steady state. Unless you perturbed the system by changing some of the input signals, everything would remain (after an initial flurry of activity, as we saw on our Irish lake) static and dull. But you would be wrong. If signals moved around a network instantaneously, then that might be so. But since signals take time to propagate, and since there are storage elements in our network (for example the stored state values inside integrators and differentiators), then most randomly wired networks would take some time to settle down, if they settled at all. Since any arbitrary network is extremely likely to contain feedback loops, the output of any one component will ultimately change its own input, and thus change its output, and hence its input…

Under certain circumstances this fidgeting will never stop, and the result is called an *oscillator* – our fifth type of building block. In electrical circuits, oscillators come in a wide variety of types, but invariably involve positive feedback and a delay. One of the simplest kinds of electronic oscillator circuit is the entertainingly named flip-flop multivibrator. In this circuit, two transistors are wired up in such a way that when one of them starts to turn on, it turns the other one off, which in turn switches the first one on even harder. On its own, this positive feedback would lock the transistors into a perpetual half nelson, but between each transistor is a differentiator circuit. Since differentiators produce an output only when their input is changing, the current stops flowing once the transistors have turned on and off as far as they are able. This releases the vice-like grip that one transistor has on the other, and the one that was originally turned on starts to turn off again. The instant this happens, the differentiator transmits the change to the other transistor, and the positive feedback drives the pair to the opposite extreme. The two transistors will continue to flip and flop from one state to the other like this indefinitely, at a precise rate determined by the design of the differentiator circuit.

An ordinary doorbell works in much the same way. When someone presses the switch and allows a current to flow, an electromagnet starts to produce a magnetic field. This field attracts the metal of a switch, which eventually opens and stops the current from flowing. The magnetic field now decays, the switch falls back to its 'on' state and the process begins all over again.

These types of oscillator are all-or-nothing systems: their outputs fluctuate from fully on to fully off and back again, and the signals they produce are known as square waves by virtue of their shape. By more subtle use of delays and mutually incompatible switches, oscillators can produce other waveforms in which the output signal varies smoothly over time and the waveform becomes triangular, sinusoidal or sawtooth shaped. Music synthesizers combine oscillators of these different kinds to produce more complex waveforms that are characteristic of natural and artificial sounds. In biological systems, oscillators provide the various rhythms of life such as the sleep cycle, the menstrual cycle and the periodic changes in cell type that take place during the development of a foetus. Both neurones and chemicals can be used to build oscillators, and we shall meet an example of a chemical oscillator when we come to build our creature's reproductive system.

One interesting variation on the oscillator is the device that electronics designers call a phase-locked loop. This is an oscillator that runs freely at its own characteristic frequency, but which can be influenced by quite subtle external influences and will readily fall into step with any similar periodic changes in its environment. Our sleep-and-wake cycle is like this. In the absence of night and day, we still show a roughly diurnal rhythm of wakefulness, but usually this rhythm is closely synchronized with the rotation of the Earth. Jet lag is the name we give to the short period of destabilization that occurs while our internal clock locks into the external synchronizing influence of daylight in a new time zone.

Another prominent biological oscillator, the female menstrual cycle, has a natural rhythm very close to that of the Moon, which seems likely to be no accident. Presumably the two have become synchronized over the millennia because moonlit nights influence patterns of human behaviour (or did before the invention of electric light – today it is easy to forget the practical reason why many important festivals are defined by the phases of the Moon, for instance). The menstrual cycle is certainly a phase-locked loop, because it is well known that women who work closely together tend to form synchronized cycles, perhaps through a sensitivity to one another's pheromones or simply through talking or noticing one another's behaviour.

Incidentally, some people might feel offended by the suggestion that their behaviour is so obviously linked to environmental influences, but the phase-locked loop is one of the most subtle and staggering of natural phenomena, and such sensitivity to the moods of the planet or of others is something to be astounded rather than offended by. Locking into natural cycles affects us all (although the female menstrual cycle is the most obvious example). It might be fashions and hemlines, styles of music, wars and famines, social revolutions or mental illness (remember the term 'lunatic'?) – if there is anything in the outside world that these behaviours can lock into sympathetic resonance with, this is what they do. Thanks largely to television advertising, we even decide to decorate our houses and buy new cars at the same time as one another nowadays. This is a testament to the subtlety and balance of our own physiological and mental systems, a sensitivity to our environment that has doubtless saved us from extinction many times in the past.

The rack, and other instruments of torture

So far we have looked at modulators, emitters and receptors, differentiators and integrators, oscillators and phase-locked loops. I could go on – take *ratchets*, for example. Ratchets are devices that drive systems forward by preventing changes from being undone once they have been made. In evolution, the ratchet of natural selection is operating on a randomly varying population, keeping all the good changes (because fitter creatures are more likely to pass on their genes to a new generation) and discarding the bad ones. In electronics, the implementation of the ratchet is called the diode: a device that allows current to flow through it in one direction but not the other. Many chemical reactions are reversible to a greater or lesser extent, but generally one end of the reaction is sufficiently more stable (lower in energy) to be considered the end product of the reaction, and so chemical reactions are ratchets too. Any device – whether electrical, chemical, physical, geographical or legal – that allows change in one direction but not the other is a ratchet. Ratchets are what drive the world forward.

Ratchets vary in type, from those that tense to those that don't. The 'rack', that ancient form of torture, is a tensing ratchet: the tighter it gets, the harder it is to turn it. The foot of a snail is a non-tensing ratchet: although it moves more easily in one direction than the other and so allows the snail to move forward, snails don't gradually get tenser and tenser, eventually to slide helplessly back to their starting point like an overstretched elastic band. Evolution presumably has some tension, because as a species becomes better fitted to its niche the number of further improvements open to it decreases. Nevertheless, because environments are in constant change and because evolutionary arms races keep upping the ante, this tension is frequently relieved and so evolution never comes to a halt. The first law of biology – adaptation – frequently lies at the heart of non-tensing ratchets: part of the system moves forward as far as it can, and then locks in place while the rest adapts slowly towards it. It is rather like trying to walk through surf – usually the best method is to walk forward when the sea is going your way, then stand firm while the surf is against you, to avoid being washed back to where you started.

Just like all the other cybernetic building blocks, the ratchet is a ubiquitous feature of natural systems, even though each individual example of the mechanism may be very different in form from the others.

Defining the bricks

The set of building blocks I have described in this chapter is not exhaustive by any means. Nor is it very cohesive – there are several levels of building block in the list, some of which can be made from combinations of others. In electronics, the basic building blocks are transistors, resistors and capacitors; things like integrators and oscillators are then made from combinations of these more primitive objects. But in biological systems some of the higher-order building blocks are so common, and turn up in such widely different contexts, that I think we need to start from a higher base than we would if we were building television sets. The Lego set I have described here is at least a good starting point from which to build complex networks of feedback.

I began this chapter by suggesting that the conventional, materialist conception of building blocks as physical things could be replaced by a more abstract one in which the material implementation was irrelevant and only the function mattered. My hope is that by looking beyond the 'parts' and focusing instead on the interactions between them we might gain some insights into the gestalt – the 'more' that is greater than the sum of those parts. Nevertheless, it turns out that for biological systems these cybernetic building blocks correspond remarkably well to a small group of physical structures and processes. Neurones are the epitome of modulators, and can be made into integrators, differentiators and oscillators. Chemical reactions can perform the same feats but in a more diffuse form. Chemoreceptors and chemoemitters capture many of the important features of transducers, and allow neurones to converse with chemistry. Genes are an excellent metaphor for describing how these structures may be combined into larger circuits, while at the same time enabling the ratchet of evolution to change the design. So, stylized versions of this small number of biological structures embody all the main cybernetic primitives. They are to organisms what transistors, resistors and capacitors are to electronics. In fact, they are more than that because, unlike electronics, they have the potential to change their configuration over time – neurones can rewire themselves as an organism learns, and chemical reaction networks can rewire themselves through evolution.

We have ended up with physical building blocks bearing biological names, yet somehow, through our more abstract, cybernetic approach to the task, we have liberated ourselves from the constraints of anatomy. I have called these elements neurones, chemoreceptors and genes because these are the names of structures that evolution has developed to perform the same tasks. But we do not have to use them in the same biological context. Our job is to replicate not the physical layout of the body but its 'informational' layout.

So at last we have our basic building blocks. They can be implemented in a computer, as first-order components – algorithms that behave approximately as if they were neurones, chemical reactions, and so on. By plugging them together into networks that solve the problems of persistence we can generate the vital spark that makes these inanimate parts into a living whole.

The last few chapters have been very analytical – we have been taking things apart to see how they are made. In a while we shall redeem ourselves by taking the other, less trodden but just as important route to nirvana: understanding things by putting them together. But first, to get us back in the spirit, I'd like to wash the dust of reductionism from our shoes by taking a short stroll through the greener pastures of holism, and reaffirm that what we are trying to do here is create something that is more than the sum of its parts.

* *

CHAPTER TEN

THE WHOLE IGUANA

Not only is there but one way of *doing* things rightly, but there is only
one way of *seeing* them, and that is, seeing the whole of them.

John Ruskin, *The Two Paths*

When I was a child I had a friend who took endless delight in pulling
the legs off spiders. I imagine the spiders didn't find his hobby quite so
amusing, but whether it was actually a cruel thing to do is rather a difficult question to answer. To me it seemed not so much cruel as a form
of desecration – the thoughtless destruction of something beautiful.
What beauty means and why it is sinful to destroy it are not what I
have in mind, however. The two observations I particularly want to
make about the antics of my arachnicidal friend are these: first, intelligence is visible, relevant and possible only through the interaction of
an organism with its environment, and all such intelligent action is
rooted in survival; and second, there is no such thing as half an
organism – life and intelligence are properties of wholes and must be
synthesized in a holistic way.

I think, therefore I (still) am

Let us begin with the first observation. Even though our hapless spider
has had all its legs amputated, it is probably still attempting to walk. Yet
with no legs to connect it to the world its actions never become realized. This illustrates something that should be self-evident: intelligence
without action is pointless. In fact, with a little qualification I would
claim that intelligence without action is not even achievable. Unless an
organism receives sensory feedback from its environment and can also

alter that environment, it cannot close the loop that makes learning and therefore intelligence possible. It is true that some creatures, most notably humans, are perfectly capable of being intelligent without interacting with the world – I like nothing better than to sit in a quiet, darkened room alone with my thoughts. But without the ability to sense and have some effect on the external world, I believe we would not be able to become intelligent in the first place.

Our internal mental world – our imagination – is quite capable of allowing us to be intelligent without interacting with the world outside. I can work out how to solve puzzles inside my head. I can even, if I so choose, live out a complete alternative dream-life in a make-believe world of my own. But I can do this only because my imaginary world is synchronized, through experience, with the real world. With no previous acquaintance with gravity, friction, pain or the effects of my actions on objects and other people, my imaginary world would be quite unformed and incapable of predicting the next chapter in its own story. For the same reason, people who are born blind cannot see pictures inside their heads – the 'rules' that govern images are unknown to them. Learning what amounts to the laws of physics through direct interaction with physical things is what stops my imaginary world being entirely arbitrary and unrepeatable. And discovering and making use of these laws of physics is what intelligence (and imagination) is for.

Intelligence is all about survival – remember the contest between the chess computer and the rabbit in Chapter 1? Intelligent systems solve problems in order to persist for longer.* In fact, even when intelligence appears to be used in a freewheeling, intellectual and inconsequential way (such as when humans play chess), it is still usually rooted in survival. Suppose I ask you a simple but abstract question, such as 'what other shape do you need in order to make octagons

---

* It is hard to avoid implying a sense of purpose in these matters. I am wrong to say that some things have intelligence and *use it in order to* survive. Evolutionary biologists have a similar problem when they find themselves saying that a species evolved a certain characteristic *in order to* solve some problem and increase its survival chances. It is much more accurate to say that creatures which evolved those features happen to survive better in the ecological niche that they occupy. They neither *choose* to evolve the features, nor *decide* to apply them to a particular problem. To put my original phrase less teleologically, I should say 'many systems that persist for extended periods do so because they solve problems intelligently'. But even that does not quite capture the truth.

tessellate?' If you choose to answer this question you are definitely engaging in intelligent activity, but you might not see that it has survival value. Yet *why* did you answer it? To exercise your skill at manipulating shapes in your head? To avoid upsetting me? To demonstrate your intellectual prowess? All of these are survival urges. The first increases your ability to react to real problems, the second decreases the risk of interpersonal conflict, and the third establishes your position and therefore your rights in the social hierarchy.

One way or another, intelligence is a mechanism for enhancing persistence, another word for which is survival. Yet many attempts to create artificial thinking machines completely fail to recognize the importance of this fact. An AI program designed to translate French into Chinese, say, gains nothing from its labours. Its actions are meaningless to it because they do not help it survive. Failure does not cause it pain; success does not bring it reward. The pleasure of translating great literature is completely lost on it. No wonder it is hard to build such devices when their activities are not grounded in anything meaningful to the systems themselves.

Many of these machines have no capacity to learn new information – they have it all programmed into them in advance by humans, so punishment and reward to teach them good from bad are an irrelevance. In my view the lack of learning ability alone disqualifies them from being described as intelligent, but even if they are counted as such they lack any incentive to act. Pain and pleasure are the driving force for all intelligent activity in the natural world and I see no reason why artificial systems should be any different. When a physical system just does what it does, regardless of whether it gains or loses by its actions, we do not normally regard it as clever. A windmill takes in corn and churns out flour, but we don't call it intelligent. Such machines have no choice but to do what they do, and what they do has no consequence for them. So how is a machine that takes in French and churns out Chinese any different?

Even if an AI system is designed to learn, the artificiality of the task and the contrived environment within which the system is obliged to exist may lead to problems. In a forced, artificial world where survival is a meaningless concept, reward and punishment are either superfluous or arbitrary. You might argue that the reward and punishment can be supplied artificially, and this is true to an extent. We can deliberately 'make life unpleasant' for the machine when it does a poor job, and give it some artificial reward when it does well. But what does 'unpleasant' mean in this context? A translation program has no emotions; it cannot feel pain or embarrassment at its failures. One of the difficulties with supplying artificial feedback to modulate the behaviour of an abstract intelligent machine is that there is little to guide us. We are unlikely to know what constitutes reward and punishment in that context, how many different kinds of reinforcement we need or when we should apply them.

The real world, complex as it is, is entirely self-consistent, and the feedback we receive from it is a natural consequence of that self-consistency. As intelligent machines ourselves, we cannot predict exactly how the real world *will* behave, but we can certainly predict how it *might* behave and how it will *probably* behave. We can rely on its essential repeatability and internal consistency to provide the data from which predictions can reliably be formed and measured. But an artificial environment comes with none of this self-consistency or meta-predictability provided for free – we have to invent it for ourselves.

This problem often leads writers of fantasy into difficulties. If the story postulates something that is apparently impossible in the real world, like faster-than-light travel or precognition, it can be very tricky to work out the implications of this new phenomenon and ensure that everything else slots into place and makes sense. Of all the possible universes, only a few seem to be self-consistent, and most fantasy stories end up having to gloss over or explain away some terrible inconsistencies brought about by the inclusion of unnatural elements in the plot. Anybody who, like me, builds artificial worlds for a living soon learns how hard it is to outdo nature when it comes to achieving self-consistency.

Incidentally, there is much talk at the moment of how intelligent systems might exist inside the Internet and serve useful purposes. Some people even speculate that the Internet could become an intelligent entity all by itself. Several hi-tech companies are currently banking their future on 'intelligent search agents' which can supposedly 'live' in the World Wide Web and find useful documents for people. But there is a real problem. The Web is a consistent world, but it is not *self*-consistent. The 'environment' of the Web consists of linked pages of text and pictures, and for the most part the text on these pages makes sense. But it makes sense to us only because we are party to extra information that exists outside the Web. Without general human knowledge about the meanings of words, and the motivations of the people who write Web pages, the lives they lead and the activities they are writing about, the Web makes no sense. Analysing it purely as a series of abstract symbols is a very difficult thing to do, and can have only limited success. In principle, an alien scientist could learn a lot about the human race by studying the symbols on the Web, but only if it already had a basic knowledge of the behaviour of physical systems and living organisms to draw on. The information on the Web makes sense only in terms of the survival value of information that lies outside it, and I think some people underestimate the relevance of this for artificial intelligence. If the Web is predictable only in terms of external knowledge, then, beyond a relatively trivial level, intelligence can work inside the Internet only if it carries that knowledge in with it. Perhaps even that is not enough, and the intelligent agents will need to have lives outside the Web too, so that the knowledge has real meaning for them.

Aside from Internet agents, some scientists are actually trying to imbue machines directly with the kind of basic knowledge that allows our own intelligence to work. For example, Douglas Lenat and his team at Cycorp in Austin, Texas, have been feeding facts and relationships into a huge computer program of a type known as a knowledge base for several years. Their hope is that this machine, called CYC, will ultimately know enough about the world to be able to solve real, common-sense problems. But why would it want to? What does all this knowledge, let alone the consequences of its own thoughts, actually *mean* to it? Without the urge to survive, what will drive it to think in the first place? I'm sure that CYC will turn out to have many applications, but I believe that something is fundamentally wrong with this kind of approach. It would be better to start with systems that are connected directly to the world (whether real or artificial), so that they are able to work out the rules for themselves.

All of the high-level knowledge we have as human beings is built on a foundation of simpler, more immediately relevant knowledge, and this in turn is grounded in our basic knowledge about how to move around in our world and survive in it. There is a hierarchy of understanding based on a chain of experience that connects us to the harsh reality of the world. Even reinforcement is hierarchical – when someone glowers at us, we automatically make a connection between this entirely harmless phenomenon, via a chain of inference, to an ultimate fear of being hurt. In our childhood, reinforcement was immediate and directly painful or pleasurable (a smack or a cuddle, say). Over the years, most of us have learned to associate stern looks with smacks and antisocial behaviour with stern looks. We behave in such a way as to minimize our risk of being smacked while maximizing our chances of being cuddled, even if nobody actually smacks or cuddles us any more. I don't see how else it could be – why should we choose not to do something, simply because we have been frowned at?

For all of these reasons, an abstract 'thinking machine' that has no axe to grind and nothing to fear, yet solves problems just because it is programmed to, seems to me to be a futile quest, or at best just a pretence at intelligence.

There is no such thing as half an organism

Which brings me to my second point. Pulling the legs off spiders does not kill them immediately, but it does seriously damage their health! Sooner or later they will run out of food and die. Dismantling any living organism, whether by amputating limbs, severing its head or even whittling it away cell by cell, eventually takes away its life, without you ever noticing where it went. As a general rule, if you take an organism to pieces you do not end up with pieces of an organism. All you get is a sticky mess of lifeless bits of meat or vegetable matter. It is possible to remove part of a creature and 'keep it alive' in tissue culture or on a life-support machine, but only by providing artificially all the systems to which it previously had access from being part of a whole. There is no such thing as half an organism. A once-living thing suddenly becomes reduced to a collection of non-living things. The life is not contained in any of those parts, but has simply ceased to exist. (Wherever I have used the word 'life' in this paragraph, you can substitute 'intelligence' and it will be equally true.)

Living creatures are not alone in this. After all, which is the most important part of a motor car? Is it the pistons? The spark plugs? The leather upholstery? In fact the question is inappropriate – a car is a complete system and it makes no sense to talk about the importance of its parts. It doesn't have a key component, because it is the *interaction* between all the components that makes it a car. It is true that some parts are more critical than others, in the sense that damage to them is more likely to destroy the 'car-ness' of the whole system, but that is not the same thing. If you start removing the parts of a car, when does it cease to be a car? Removing the seats doesn't stop it being a car, nor for that matter does removing the spark plugs – it will not work without them but it is still a car. If you chip away at the cylinder block and shave bits off the bodywork, sooner or later you will come to a point when you realize you are no longer looking at a motor vehicle. But this critical threshold will always somehow lie in the past – you realize that the car-ness has gone, but you didn't notice it go.

Intelligence is like that too. We could start whittling away at the brain to see which part of human intelligence is the 'most important'. We can remove the visual cortex – we know that people can be blind and still intelligent, so that bit can clearly go. Similarly we can cut out the auditory cortex, because deaf people can still think perfectly well. What about our sense of responsibility and ability to make plans? People with damage to their frontal lobes often act irresponsibly and antisocially, but they can still reason and respond to their world, so a prefrontal lobotomy would not seem out of the question. Memory is a bit trickier: I manage to get by with a very poor memory for names and facts, so this cannot be a critical faculty. A very few people lose almost all their ability to lay down or recall memories, and although their lives are severely handicapped, they still manage to interact with the world and survive. A brain with absolutely no facility to record the past is one that cannot learn, so it cannot become intelligent. But we can remove individual *types* of memory – the kind we use to remember faces, or the kind that enables us to remember how to ride a bike – without the whole system collapsing between our ears. Language can go, too – it may be quintessentially human, but people are still people, even when they have suffered damage to the parts of the brain that handle the interpretation or generation of language. What is more, Mr Spock and Data from *Star Trek* seem to manage perfectly well without emotions, so the emotional parts of our brains cannot be necessary, either.

In short, we can remove any single aspect of intelligence without destroying the whole. But if we start removing these faculties one by one, sooner or later what is left will cease to look intelligent. There will be a point at which the removal of a single component causes us to cross a subjective threshold between intelligent living thing and lump of meat, but if we start again and remove the parts in a different sequence, it will be a different component's removal that represents the last straw next time around. Similarly, if we take these individual faculties and start combining them, there will come a point at which it seems reasonable to describe the new entity as intelligent, but we would not ascribe intelligence to any of the faculties individually.

So why do so many who attempt to create thinking machines expect to be able to implement just one specific aspect of intelligence and get away with it? You may think I have laboured this point quite heavily, but I think I have reason to – most attempts at artificial intelligence over the past fifty years have tried to construct just one faculty quite independently of the others. Most often, the chosen faculty has been the ability to reason logically, but sometimes it has been the ability to recognize patterns (whether from a visual scene or spotting trends in the stock market). Very rarely indeed has anyone tried to implement several of these faculties at once in order to build an integrated whole.

Perhaps some researchers would argue that building the whole shebang is just too difficult, which is why they have concentrated on the more tractable task of building some of the individual components, perhaps with an eye to fitting them together someday. I can quite understand their fear about the enormity of the undertaking, but I think their solution is misguided. For a start, how do they know that the bits they have built can be fitted together at all? It may be relatively straightforward to attach a computer vision system based on neural networks to an inference engine based on rules and facts (but then again it may not!), but can they then incorporate language into the system, or the ability to coordinate movement? Can a learning algorithm be added to an expert system designed by someone else? Since all the bits are designed at different levels of description, with different techniques and in ignorance of one another, fitting them together into a unified system is not as easy as it looks.

Attempting to synthesize the individual characteristics of intelligence separately and then slotting them together is essentially the same 'outside-in' approach that I earlier criticized in relation to virtual-reality simulations. There is a danger that this outside-in approach will break the problem down in the wrong way. To take a random illustration: an elephant has skin like parchment, a trunk like a serpent and makes a noise like a foghorn, but wrapping a cobra round a lighthouse containing the Magna Carta does not give you an elephant.

What is more, many of these intelligent faculties may not even be possible without the simultaneous existence of the others. We do not have a separate part of our brain that handles logical reasoning, for instance. There may be areas of brain that are more deeply implicated in this particular faculty, but they rely heavily on many other parts, perhaps *all* the other parts, in order to function. For example, I happen to believe that our ability to reason consciously is grounded in our ability to build mental models of the world, and that this in turn is founded upon a biological mechanism for constructing predictive models of body movements and simple sensory hypotheses (such as the automatic ability to 'fill in' the missing details from a partially obscured image of a face). Perhaps this mechanism originated to handle the more rapid and complex (but not necessarily very intelligent) movements required by animals as they evolved from living in the sea to living on land, and our own higher mental faculties are an emergent consequence of a system that grew upon and made use of this more primitive foundation. Similarly, our emotions are products of the basic drive mechanisms that evolved to control the behaviour of relatively primitive organisms such as fish and amphibians, and now form a crucial part of the value system that motivates our actions and with which we measure our experiences. Fear and anger are probably relatively simple, built-in mechanisms, while grief and embarrassment are perhaps more subtle shades mixed from these primary colours.

So imagine a situation in which someone rushes out into the street to save a child from oncoming traffic. They are certainly employing intelligence. But their action is based on a prediction of what might happen next that probably evolved for handling rather low-level plans and forecasts, plus an emotional response to the situation derived from the mechanism that evolved to modulate feeding and fleeing behaviour. Their visual pattern recognition enabled them to see the car and the child; their memory threw up images of car accidents and reminded them of how it feels to be struck by something heavy. Their planning and navigation circuits assembled a plan of action, and their motor sequencing system carried it out. No single part of their brains did the thinking – it was an emergent consequence of the interactions between all these parts.

In a few months' time I shall be going to a small workshop in Lanzarote. Among the attendees will be Daniel Dennett, who coined the phrase 'the whole iguana' that I used as the title for this chapter. It is also the title of the workshop, and all the delegates will be people who specialize in building or understanding whole living systems. A couple of years ago I co-organized a workshop in Cambridge, UK, also on the topic of complete artificial living things (this workshop gave me the impetus and opportunity to write the book you are reading now). For the two workshops we have managed to muster about a dozen (admittedly high-calibre) speakers who deliberately take an inte-grated-systems approach to their work – a dozen scientists from a pop-ulation of many thousands of AI researchers. Yet go to a conference on case-based reasoning, or computer vision, or genetic algorithms or any of the many other piecemeal techniques, and you will find large numbers of delegates who study these subjects in minute detail, to the exclusion of all else. People who attend computer vision conferences rarely go to meetings about automatic translation, and delegates at those are rarely to be found at workshops on neural networks. They don't even speak the same language or share many concepts. As usual, reductionism rules, and nobody can see the wood for all the trees.

A specification for a thinking machine

So when we pick up our cybernetic tools and start to build a machine with a mind of its own, don't be surprised to find me talking about such apparently unrelated matters as digestion and the immune system. These are both parts of the whole. Neither is crucial, but the thing we are trying to construct has to be made from at least a few such things. Having a reproductive system may not seem critical for the emergence of intelligence in an artificial world, and it is not. Yet a reproductive system is a computational mechanism that provides some of the moti-vational factors and a part of the value system for initiating and regu-lating intelligent activity. This is going to be a whole organism. It won't be a very bright one – don't expect it to translate French into Chinese or diagnose diseases. Don't even expect it to make plans or use tools. Nevertheless, it will have the following characteristics, many of which depend on the interactions of the whole, integrated system.

First, it will have a brain, made from neurones and biochemicals. This brain will be able to learn for itself, without assistance, in a fairly complex, realistic environment. It will be able to recognize situations it has met before and learn how to respond to them. It will also be able to generalize from past, similar situations to help it react sensibly to novel circumstances. As well as being able to remember, it will be able to forget – nerve connections will be recycled, so that the creature retains only important things. The brain will handle basic control of attention, bringing important events to the creature's notice so that it can focus its senses and actions on the most salient objects.

Second, underlying the brain will be a simple emotional system of drives and needs which motivates the creature to act and regulates its learning. This mechanism will be deeply interconnected with the other bodily systems such as digestion, linking brain inextricably with body.

Third, the creature will need to eat to stay alive, so its intelligence will be grounded in survival. In addition, it will be susceptible to disease and have a simple immune system that responds to infection. It will also have to cope with other objects in the world, including other creatures. Through these things, the creature's survival depends on its ability to interact with its environment and learn from it.

Fourth, it will have a primitive ability to speak and understand language, enabling it to communicate with its owner (you). You will be able to give it simple commands and teach it about its world, and it will be able to tell you what it is doing and how it feels.

Last but not least, it will have a comprehensive reproductive system which enables it to attract a mate, conceive and give birth. Its entire neural and chemical structure will be encoded into genes, which it will pass on to its offspring. Children will therefore be built to a design based on that of both their parents. Sufficient features of their reproductive system and genetics will exist to enable the race to evolve over time, including the evolution of complete new chemical and neural structures.

That will do to be going on with. We are about to create a high-order persistent phenomenon – a tangled web of feedback loops that has the ability to persist over an extended period in both of the ways we require. Intelligence will enable it to learn and adapt to its environment, and so persist as an individual. Genetics and the appropriate mating behaviour will enable the recipe for the pattern to persist from generation to generation.

Let's create life!

CHAPTER ELEVEN

* *

IGOR, HAND ME THAT SCREWDRIVER …

I beheld the wretch – the miserable monster whom I had created.

Mary Shelley, *Frankenstein*

You have been very patient, and at last the time has come. We have laid out our instruments, prepared our potions and donned our sterile masks. Outside the rain-lashed window, lightning struts among the Gothic turrets as Thor's reproachful voice grumbles ominously at us from beyond the mountains. While we wait in apprehension for the giant capacitors to charge up with life-giving electricity, we have time for one last pre-operative briefing to recap the main principles behind what we are doing.

Dr Frankenstein's creed

Life is not the stuff of which it is made – it is an emergent property of the aggregate arrangement of that stuff. Even the stuff itself is no more than an emergent property of a still smaller whirlpool of interactions. Living beings are high-order persistent phenomena, which endure through intelligent interaction with their environment. This intelligence is a product of multiple layers of feedback. An organism is therefore a localized network of feedback loops that ensures its own continuation.

Intelligence cannot be abstracted – we have to build a whole organism. Neither can intelligence exist in a vacuum – it has to be embedded in a self-consistent environment. Life is the sum total of all the feedback within the organism, and between the organism and its environment. The division between organism and environment is not a real boundary, but a convenience dreamt up by our own brains – the universe is really just a single jumble of interactions.

A computer cannot be intelligent or alive. Nor can a computer program. But a computer can be used to create a cyberspace. Inside that cyberspace we can construct first-order objects and use algorithms to emulate their behaviour. These objects are not alive or intelligent either, but they can be pieced together to build a second-order assemblage that is. Our task is not to program in intelligent behaviour, but to enable such behaviour to emerge from simulated objects that embody the cybernetic properties from which life emerged in the natural world.

To complete the picture, we must ensure that the recipe for this emergent phenomenon is not hard-wired but is able to be passed on from generation to generation and modify itself in order to persist on longer timescales, as the environment changes. Our creature will be fully alive and intelligent only if its future lies in its own hands, and to give it this autonomy we must relinquish direct control of its design. In short, the plans for how to assemble our creature should be coded in its genes.

Now that we have recited the creed it is time to put it into practice. 'Scalpel, please, nurse ...'

Welcome to my world

Assume that we already have a program that simulates a virtual environment for our creatures to exist in. We have defined the properties of space, and we have programmed in factors such as gravity, as described in Chapter 6. The world is filled with virtual objects – toys, food, vehicles and assorted other items (even simulated balls). Each of these objects is controlled autonomously by its own short program 'scripts' which define how it is to change its state and appearance as a result of 'messages' sent to it from other objects in the virtual world. One object can trigger an action in another object by sending it a message, and these messages are transmitted over short or long ranges, through barriers or not, as appropriate, in order to simulate the transmission of information by touch, sight or sound. The heart of our computer program is a time-slicing loop, which gives each virtual object a frequent and regular opportunity to update itself by carrying out an instruction or two from any of its currently triggered scripts.

We now have the wherewithal to construct a rich and interesting (and dangerous and taxing) playground for our creatures to live in. By defining a series of message types and programming our objects to respond to these messages and emit new ones, we can make a world that looks very much like the real one. Objects trigger other objects, which change their state and appearance and trigger other objects. By the magic of feedback and with a good deal of artistry, it is possible to create a complete ecosystem of interacting structures that never quite settles down and never becomes boring, for us or for our creations (see Figure 16).

Our virtual world can be as complex and rich as we like, because one of the immense advantages of this approach to computing (the technical term for which is 'event-driven object orientation') is that we only have to think about the properties of one object at a time. We never need to understand or attempt to control the behaviour of the entire world as a single entity. This is in stark contrast to the traditional approach to software (not to mention most other types of endeavour), where a huge monolithic structure is designed from the top down.

Figure 16. Part of the *Creatures* world.
© CyberLife Technology Ltd 1997

Working from the bottom up is a skill that takes some acquiring, but it pays dividends, especially when the thing you are trying to create is a population of complete thinking creatures, embedded in a rich, surprising virtual world. Trying to look at the whole thing from above and retain control of it is totally futile, so we might as well let go gracefully right from the start.

I really cannot stress this attitude of mind too strongly – instead of 'command and control', we need to 'nudge and cajole' (I must print this on a T-shirt sometime). Whether you run a school, run a country, manage an ecosystem or write computer software it makes no difference: complex adaptive systems cannot be dictated to – you have to learn how to go with the flow and nudge individual components in order to encourage the system to go in the direction you want it to. To do this you need to understand feedback and how to prevent complexity explosions – the exponential growth in the number of things you need to know and manage as the system gets larger. Nature is a real expert at this, and everybody would benefit from a study of biology to see how the trick is done. And by 'biology' I mean the study of complex adaptive systems, not the 'naming of parts' that often passes for biology in schools – a good biologist studies the commonality between things, not their differences.

The monster takes shape

By using programming techniques similar to those I've just described, we can also create a style of virtual object that will become the body of our creature. We could perhaps assemble this from a collection of simpler autonomous objects (looking like forearms, heads, hands, and so on), linked to one another by defined angles. We could then program them so that when the angles between any of the joints are adjusted, the whole collection of parts will be redisplayed in the correct new position. By manipulating the angles between leg joints we can raise our simulated body's foot from the floor and move it forwards. As the foot collides with the floor again, our mathematics will detect this in much the same way as when we triggered the bouncing of a simulated ball earlier. Attempting to push one foot down through the floor will actually cause the body to rise and the other foot to lift, and so by a carefully choreographed sequence of rotations of its body parts, our creature will take its first faltering steps.

None of the programming that I have mentioned so far is trivial, of course, and a lot of effort goes into creating such a virtual world. But although a comprehensive, flexible, self-consistent virtual world is a vital necessity, we do not need to consider how to create it in any more detail here. Our main task lies at a higher level – we have to define how other such virtual objects should be combined together and allowed to communicate with one another in order to build a structure with something approaching a mind of its own.

Since our creature is starting to take shape and can move like a living thing, we might as well give it a name. It is really rather early to succumb to anthropomorphism about what is still only a collection of vectors and graphics code, but if nothing else it will save me having to refer to 'our creature' so often. I am going to call him Ron, after the very first of these creatures that I built. I admit that Ron does not sound a very heroic name to choose for the progenitor of an entire race of artificial life forms. Nevertheless, you might be interested to know that the fabled King Arthur, leader of the Britons and wielder of the mighty sword Excalibur, also happened to own a spear called Ron. For some reason we hear a lot about Excalibur, but never much about Arthur's exploits with Ron. Also, the name reminds me of a series of medical articles in *Reader's Digest* that I read as a child. Each article looked at an organ of the body from its own perspective, under titles such as 'I am Jane's spleen'. I may be completely wrong, but I seem to recall that one of them was headed 'I am Ron's brain.'

Making sense of things

We have defined Ron's body as a collection of linked body parts, and by controlling the angles between them we can make him move, walk, run, crawl or dance. We can even make him approach another virtual object and activate it by sending it a message, so Ron is capable of having an influence on the world by pressing buttons or hitting things (Figure 17). Perhaps we can also send objects a 'grab' message which causes them to link themselves temporarily into Ron's body-part collection and become 'carried' by his hand. So far, so good. By executing

Figure 17. Ron meets a virtual ball.

short program scripts stored along with the creature object, we can do all of this and more. Our virtual Frankenstein's monster can now manipulate his environment. But we also need him to be able to sense the state of his world.

By simple trigonometry we can work out which objects Ron should be able to see or hear from his present position, given the direction of his gaze. So when objects emit messages to say that they have changed state in some way, we can decide which of these messages can be heard, seen or felt by the creature. We can also scan the local environment in a more active way to determine other information about objects' status. How far is Ron from a wall? Is the object he is looking at moving towards him or away from him? How quickly? What sort of object is it? All these pieces of information can be encoded as if they were electrical signals, ready to be pumped into specific neurones in Ron's brain.

By such means we can give our creature a sensory knowledge of its external world, and we can limit that knowledge in realistic ways so that things cannot be seen if they are not looked at. We can also give Ron senses that respond to his *internal* environment. What is the current disposition of his limbs? Is he moving forward? Is he tired? Many of these internal senses are not updated by program code but by the creature's own biochemistry, more of which later.

So Ron has a rich array of information about his sensory state, and a rich repertoire of actions that he can perform on the external world. But what connects the one with the other? The answer, of course, is that Ron has to have a brain whose task is to filter and interpret the torrent of incoming sensory information, use it to work out a sensible (self-preserving) response and trigger appropriate actions.

Ideally, Ron's brain should also be capable of at least two other complex tasks. First, it would be able to take raw sensory data such as images from the eye or sound waves from the ear and process them to create more abstract, high-level information in order to recognize objects, extract motion information, and so on. Second, it should be able to carry out the coordination of limb movements all by itself, so as to execute complex and context-dependent sequences of motor actions. Moreover, it would be able to learn how to do both of these things for itself. Each newborn creature would have to learn how to recognize a carrot from any angle, and to associate carrots with other foodstuffs to form a self-defined category of edible things. It would be able to learn for itself how to walk, and how to combine walking with reaching in order to pick up the carrot and eat it.

But I think we have quite enough to worry about without this. Happily, one of the virtues of a virtual world, compared with the real one, is that we can cheat. We do not need our creature to be able to recognize an object by its shape, for example. The program knows perfectly well what kind of object it is, so it can present this information directly to the creature's senses. Similarly, we can define the correct sequence of movements for walking and so on in advance, and code them into program scripts, rather than find a way for interacting neurones in the creature's brain to encode and enact such sequences.

This is an embarrassing cheat, I know. When I wrote the *Creatures* game (on which this chapter is based), it was a decision that I felt was justified, given the number of other problems I had to deal with and the limited computer power available for processing it all. Since then I have been working on ways to encode these extra facilities inside artificial brains, and future creatures that I build will have these more rounded characteristics (especially the robots, where cheating is not so easy). They will also have far more sophisticated 'thinking' abilities than poor, dim-witted Ron.

Even though this is an admission of defeat, we can make up for it a little by encoding some of this information in genetic form. Since the body posture of a creature is defined by a list of angles, and an action is defined by a sequence of target postures, we can encode all the creature's motor 'skills' into its genes. An individual creature may not be able to learn to walk or modify its movements in the light of experience, but through natural selection the *species* as a whole can evolve new and better motor actions, improving on the ones we have programmed into the initial generation.

So we can ignore the visual recognition and motor sequencing aspects of mental function. The primary task our artificial brain has to deal with is learning for itself how to connect these predigested sensory inputs to the appropriate, equally predigested motor outputs, and so cause the creature to respond sensibly to changes in its environment. There is also one other task this little brain must be able to carry out, and since it provides a good introduction to the properties of our simulated neurones, we shall look at this much simpler problem first.

Pay careful attention!

In such a rich environment filled with active objects, Ron is already in severe danger of information overload. Not only might there be many objects in view or earshot at one time, but also each object is generating several items of sensory information. Ron might simultaneously be able to see an approaching elevator and another creature moving away, while hearing a radio making a noise. How does he know which one is moving which way, or whether the 'I can hear a noise' sensory signal relates to the radio or the elevator? This is a problem known as sensor fusion – how to take various sensory streams and relate them to one another and to specific sources. Ron cannot have a separate set of sensory inputs for every single object of which he might be aware because the number of input lines, and hence the amount of work his brain has to do, would be enormous.

So how does nature tackle this problem? One important solution is the use of an *attention* mechanism. Many animals (including humans) have a kind of built-in tunnel vision which obliges them to a greater or lesser degree to attend to one thing at a time. We do this partly by physically controlling our sensors – we direct our gaze and focus our sight on a single object at a time – and partly inside our brains by filter-ing out the bulk of the sensory data that are assailing us at any one moment.

Everyone has come across the 'cocktail party effect', in which we are able to attend selectively to a single conversation in a room full of people talking. This is really quite a sophisticated neural mechanism. I have never been very good at singling out a conversation in a noisy environment myself, but yesterday I was marvelling at another inter-nal searchlight that we all share. I was stopped at a traffic signal, but had positioned my car badly and my rear view mirror was partially obstructing my view of it. With my right eye I could see the signal, but with my left all I could see was the reflection of the motorcyclist behind me. I quickly discovered that I could select which of these objects to 'see' at will, without moving my eyes. Traffic light, motorcyclist, traffic light, motorcyclist … It took about a second to flick from one to the other but the effect was really quite dramatic as one object appeared and the other vanished. Deliberate control of normally automatic mechanisms is not always a good idea, though, because I became so engrossed in this experiment that when the lights changed my mind failed to register the fact. The gesture made at me by the motorcyclist was brought to my attention rather more quickly.

One of the features of this 'internal gaze' is its ability to select the most *important* sensory signal and bring that signal to our attention if necessary. Our sensory filters do not block everything out completely – large parts of our brains are still monitoring what is going on and inter-preting its significance – it is just that our brains don't trouble our minds with it all. That is, unless priorities change. If we are listening intently to the TV and the clock is ticking, we are completely unaware of the sound of the clock. However, if the clock stops our brains sit up and take notice. Something has changed and it might be important, so our attention is drawn to the change and we shift our gaze in the direc-tion of the clock.

We can implement something like this in our artificial creatures sur-prisingly easily. True, it will be far less sophisticated and powerful than our own internal attention mechanism, but it is better than nothing.

Suppose we feed a signal into Ron's brain every time an object changes state in a way that his senses can detect. If something changes in a visual way and he is looking in its direction, or in an audible way while it is within earshot, or in a tactile way when Ron is touching it, that information is sent to his brain. All we need to send is the intensity of the stimulus and the identity of the object that caused it. The exact nature of the change will be picked up by Ron's primary senses if he chooses to shift his attention in response to the stimulus, otherwise he will simply ignore it.

Assume that the world has a limited classification of object types, so that for example carrots, apples and honey are all 'food' as far as Ron is concerned. If each category of object is given its own input wire into the brain, and the intensity of the signal on each wire is proportional to the change in a nearby object of that type, then the brain can easily work out which object to concentrate on. Inputs relating to nearby objects that are changing rapidly will carry frequent signals; those that are changing dramatically will carry strong ones. At any one moment the object that has been making the most 'fuss' (changing rapidly, dramatically or both) is the object to which Ron should pay attention. The ability to combine changes in both intensity and frequency into a single signal is a property we have already met in one of our cybernetic building blocks: all we have to do is *integrate* these input signals over time.

Neurones make great integrators – we simply charge them up (increase the value of their internal state) in direct proportion to the strength of their input signals, while simultaneously letting them discharge exponentially back towards a zero rest state. Each input impulse will pep up the level of activity a bit, but the decay process will start to lower it again. For any given series of input signals there will be an activity level at which inflow is balanced by outflow and the neurone settles down to a constant output value, which is a measure of the average amount of 'fuss' an object has lately been making.

By connecting the input signals caused by changes in nearby objects to an array of leaky integrator neurones in such a way that each class of object is represented by a different neurone, we can generate a sort of bar chart of the current amount of commotion in Ron's environment. The bar that is tallest (in other words, the neurone that is firing most strongly) represents the type of object making the most detectable fuss (perhaps signalling a source of danger or amusement), and so this is the object to which Ron should probably pay the most attention. We can work out which neurone is firing most strongly by using a method called winner-takes-all. If a neurone starts to produce an output signal, we feed some of this signal back to the neighbouring neurones in such a way that it tends to suppress their own tendency to fire. Each neurone is trying to fire, and each is simultaneously suppressing the others. As soon as one starts to get an edge it will gain more suppressive power over its neighbours and hence be proportionately even stronger than them. Very quickly, one neurone will fire strongly and suppress all the others completely. This, of course, is positive feedback in action – to them that hath is given more.

All we then have to do is to couple the creature's neck to this collection of neurones, so that it looks in the direction of the object that is currently winning the race. Ron will now automatically pay attention to the object making the most fuss (which may include objects that have simply come into view for the first time). By swivelling his gaze, his other senses will become focused onto this one object and the other parts of his brain can then detect more detailed information about it, such as what it is and why exactly it is making such a commotion. The bulk of Ron's brain need pay no attention at all to the other objects in the vicinity unless they start to do something more significant. So Ron might happily ignore a sleeping tiger and instead pay attention to a toy that has come into view which he has never seen before. The novelty value of the toy outweighs the danger from the tiger for the moment, but if the tiger stirs it will quickly grab Ron's attention.

There is an added bonus from this simple mechanism. Ron is aware of only one object at a time and gets information about that object alone, so when he makes a decision to perform some action we can reasonably assume that this action (if it is not an 'objectless' action such as 'walk east') should be performed on that same object. We can refer to the current object of Ron's attention as '*it*', because we can use it in sensory statements such as '*it* is moving towards me' and also motor statements such as 'run away from *it*'.

And so we have built a very simple part of our creature's brain. One region of neurones performs the specific task of controlling attention. This was a really trivial structure to build, but it shows how we can use the properties of neurones (their ability to sum signals, their tendency to 'leak' charge and their facility for suppressing one another) to perform a simple computation. Having chosen which object is to be *it*, and established what *it* is up to, we now have to do the tricky bit, which is to decide how to react to changes in *it* or the rest of the external or internal environment, and how to learn from experience which are the best options to take in any circumstance. We need to give Ron the ability to make his own decisions.

* *

CHAPTER TWELVE

I AM RON'S BRAIN

Pooh began to feel a little more comfortable, because when you are a
Bear of Very Little Brain, and you Think of Things, you find sometimes
that a Thing which seemed very Thingish inside you is quite different
when it gets out into the open and has other people looking at it.

A.A. Milne, *The House at Pooh Corner*

The human brain is a mind-bogglingly powerful device. Not only can
we assess which are the most important features in our environment
and pay attention to them, but we can learn to recognize them, group
them into categories, imagine them in their absence, predict their
likely future behaviour, form plans about them, act out those plans by
coordinating hundreds of muscles, reason about the behaviour of
abstract (objectless) properties, form analogies and extrapolate from
them, and, last but not least, be aware that we are doing it all and thus
be conscious of our own existence. Copying the functionality of the
whole human brain is really rather a tall order, not only for this book
to relate but also for a digital computer to simulate and for human
ingenuity to decipher. But we can certainly aim to get as many properties of brains as possible to emerge from as simple a mechanism as we
can get away with. Seeking an elegant minimal structure from which
to obtain maximally lifelike behaviour was the balancing act that was
uppermost in my mind when I started designing an artificial life form
in earnest, back in 1992.

My task was to write a commercial computer game in which creatures such as Ron would star. I knew I had limited time in which to do
it, and also that I had to produce something that really worked, and not
a brilliantly argued theoretical paper explaining why I'd failed to
deliver anything. All of this focused my mind, and it became a matter

of urgency to work out a practical way of turning a pile of simulated neurones into a working brain. After some weeks of prevarication I decided to do what I normally do on occasions like this. Like Moses, I went up onto the mountain to think, and refused to come down until I had cracked the problem.

So imagine we are sitting on a dewy, green hillside, 250 metres above sea level, staring out across a large part of the south-west peninsula of England. To our right glistens the muddy water of the Bristol Channel, and ahead are the low-lying marshlands of Somerset, dotted with hills that were once islands and seem to be so again as they peek above the early morning mist. The marshes were drained in medieval times by the monks whose abbey still lies at the foot of myth-ridden Glastonbury Tor, which we can see off to our left. Glastonbury is strongly associated with King Arthur (who, you will recall, had a spear called Ron), so this is a great place for us to sit and contemplate a design for Ron's brain.

I wanted to set the scene because, although the story that is to follow may look pretty straightforward and logical, designing new things is much more of a creative, artistic act than the following chain of deductive logic might make it seem. When Moses went up the mountain to think, he only had to come back with a few rules for good behaviour, but I had to devise a method for thinking my way towards the design for an intelligent being. I needed my creative muse with me, and she seems to prefer hilltops.

After a couple of days getting nowhere, I finally hit upon the following line of reasoning: if I don't know how to design a brain that can learn to solve problems for itself, perhaps I could imagine what one would look like *after* it had solved all the problems it could ever face. By looking at the destination (however unattainable), perhaps I could glean some clues about the route. In other words, what would an infinitely experienced creature look like?

Believing in sky-hooks

If an architect believes for a moment that there are hooks in the sky to hang his creation from, he may be able to conceive of structures that he would otherwise not dare to think about. Once the design starts to take shape, he may then begin to see ways in which the essence of it can still be achieved without the need for sky-hooks at all. Maybe this will work for us, too. An infinitely knowledgeable creature – never mind for the moment how impossible such a thing might be – would presumably be one that knows exactly what to do in every conceivable circumstance. If so, it would have to fulfil just two requirements: first it must be able to distinguish all those circumstances from one another and then, when it has recognized which situation it is in, that fact alone should tell it what to do.

This sounds like a really easy thing to wire up (Figure 18)! Any individual circumstance is defined by its unique combination of features. A large green thing coming towards me and a small red one moving away are clearly different circumstances. It is easy to imagine how neurones can encode such truths: we can simply connect any one neurone to all the sensory inputs and cause it to fire when the inputs form a certain pattern. More efficiently, we can connect that one neurone only to the sensory inputs that represent the features we want it to recognize, and cause it to fire when all its inputs are firing simultaneously. In digital logic this is called an AND gate, because it switches on only when input one AND input two (etc.) are on. We

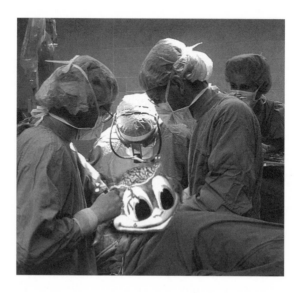

Figure 18. Brain surgery for beginners.

would really like something rather less all-or-nothing, so that we can also detect *how much* each feature is represented, as well as whether it is true (the importance of this will become apparent later).

So, recognizing every conceivable circumstance is in principle an easy thing to do, using a neurone to represent each possible pattern of inputs. Deciding on the correct action to take in response is even easier, because we can simply wire up the output of each 'recognizer neurone' to the appropriate 'action neurone'. If a big green thing starts to move towards our perfect creature, the neurone that recognizes the pattern 'it is green AND it is moving towards me' will begin to fire. The greener it is and the faster it is approaching, the more intensely the neurone will fire. If we have assumed that green things are dangerous, then this neurone will already be wired up to send a signal to the 'run away from it' neurone and increase the chance that it too will fire. The action neurone that fires (or the one that fires the strongest, if several courses of action are possible) will trigger the appropriate behaviour. Easy!

But this is a hypothetical perfect creature, and perfect creatures do not exist. For one thing, the number of possible situations in which even our simple creatures can find themselves is absolutely enormous. Given 128 sensory inputs (the number I coded into my creatures), and assuming each of them can simply be on or off, the total number of input patterns is 2^{128}, or roughly 3×10^{38}. We would need as many neurones as there are atoms in a thousand million tonnes of hydrogen gas! Representing every possible sensory situation would clearly require rather more neurones than the average personal computer can simulate.

The other slight snag is that poor Ron does not have a clue about which is the best course of action to take in every conceivable circumstance – we need him to learn this for himself, by trial and error. But trial and error alone is not good enough, because a creature with, say, sixteen different possible responses to any new circumstance, is going to make rather a lot of errors before it learns the best response. At best this would look pretty unlifelike, but worse still, some decisions in life can only be made once, because mistakes can have fatal consequences.

But thinking what an infinitely wise creature might look like has certainly given us a plausible basic architecture we can work from – a series of pattern-recognizing neurones connected to a smaller series of action neurones, which in turn are connected together in a winner-takes-all fashion. Now all we have to do is address three problems: we lack the resources to record all possible circumstances, we lack prior knowledge about the best actions to take, and our creature will lack the ability to face up to novel circumstances with anything better than a one-in-sixteen chance of choosing the right response.

Memories are made of this

Our basic architecture is capable of encoding two kinds of relationship. It can store a set of relationships between features which it uses to recognize the current situation. It can also record the relationships between these situations and their appropriate actions – what we might describe as the rules for behaviour. But since the creature cannot know all the possible circumstances and their correct responses in advance, it will have to build up *memories* of these two kinds of relationship as it experiences or discovers them.

One thing we can safely say is that although our creatures could potentially find themselves in zillions of possible different circumstances, in practice any individual creature will experience only a tiny subset of those circumstances in its lifetime. Of course, each individual will encounter a rather different subset. So if we were to judge that the total number of circumstances a single creature will *actually* encounter in its lifetime is sufficiently small, we can simply add new memories as those circumstances crop up. In other words, we can allocate a finite (but large) number of initially unconnected or randomly connected recognizer neurones (I called them concept neurones, but 'percept' would have been far more accurate). We can then allow them to *wire themselves up* to represent new situations, as and when the creature encounters them.

This is not quite as easy as it might seem. For various reasons we want to stick to a bottom-up approach, so each neurone must be able to decide for itself when and what to connect to, in the absence of any direct global knowledge of what the other neurones are doing. If we are not careful, as soon as the creature has its first sensory experience all the unallocated concept neurones will shout, 'Me! Me! I'll do it!'

and every single neurone will encode the same memory. I put quite a lot of thought into solving this and still wasn't happy with my solution. I admit to having cheated a bit!

In fact, even this idea of wiring up another memory every time something new happens will not do. This is because the number of circumstances that even a single creature will encounter during its lifetime is still staggeringly great, and its brain would have to be many orders of magnitude larger than could be simulated using a desktop computer. Perhaps we can still lay down new memories as experiences occur, but also add a mechanism for *forgetting* those that turn out not to have been important. As it happens, this can just work well enough within the few hundred neurones we can afford. So now we need a mechanism for remembering and another mechanism for forgetting, plus some rules for deciding which situations to remember and which to forget.

Even more efficiency can be gained if we do not connect every concept neurone to every sensory input. Perhaps only singles, pairs and triplets of sensory signals ever need to be remembered in combination – the creature might need to distinguish between 'something is moving towards me' and 'something edible is moving towards me', but will probably gain nothing from memorizing the fact that 'it was edible and approaching *and* green *and* it was Tuesday *and* I was bored'.

So we can define a large but not astronomical set of concept neurones, each of which has only one, two or three input wires (dendrites) and each of which is initially loosely connected to a random set of sensory inputs. Over time, these connections will break loose (be forgotten) and form anew (memorize a new situation), perhaps only to break loose again later on when the situation turns out to be not worth remembering after all. In other words, we can recycle our neurones.

The other kind of relationship our brain has to learn is the one between a situation and an action. This can be memorized in a very similar way. Each of the action neurones (one per possible action) can have a large number of input dendrites, each capable of seeking out and connecting to a concept neurone to form a memory of a relationship between a situation and a given action. By making these links only between concept neurones that are actually firing (the present situation) and action neurones that are also firing (the action being taken), we remember only relationships that actually occur, and not all possible condition/action pairs.

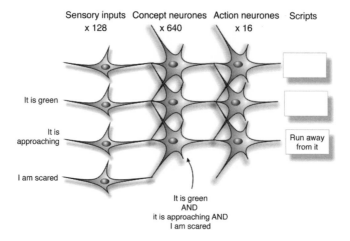

Figure 19. A slice of brain.

Thinking a thought

So we can imagine a system with two kinds of neurone: large numbers of concept neurones, each of which has a small number of inputs, and small numbers of action neurones, each of which has a large number of inputs. Both kinds actively seek out new connections and lose old connections (Figure 19). We shall look at the rules for remembering and forgetting when we cover the biochemistry of reinforcement in the next chapter. But there are two more things we need to add to our model to make it work properly. One is 'approval/disapproval' and the other is 'generalization'.

We shall start with disapproval. Simply connecting a concept neurone (representing a sensory situation) to an action neurone (representing an appropriate response) is not the whole story. If signals from concept neurones increase the activity level of the action neurones to which they are connected (these action neurones are leaky integrators, like the ones we used in the attention mechanism), then when a creature finds itself in a given this situation, certain concept neurones will fire and effectively *recommend* a particular action to the creature, by increasing the chance that an action neurone will win the winner-takes-all and cause its associated action script to execute. So far, so good. But what if that action turns out to be a really stupid thing to do?

Most lessons in life do not teach you what to do – they teach you what *not* to do. Don't tell your boss at dinner that she has had spinach on her tooth all evening; don't try to tickle tigers. So we need concept–action relationships that can *disapprove* of actions as well as recommend them. If an action turned out to be advantageous in a given situation, we must ensure that this action is more likely to be carried out in the future, which we do by making a connection between condition and action that excites the action neurone whenever that condition is true. On the other hand, if in a given situation an action turned out to be a stupid decision, we must make sure that in future that situation discourages that action, by tending to *inhibit* the firing of the action neurone. This is not difficult to do – we just design our simulated neurones in such a way that the signal coming into them can be amplified by an amount that can be negative as well as positive. So when an action proves beneficial, we make any relevant connections more excitatory; and when an action proves to be a bad move we make the connections progressively less excitatory until ultimately they become inhibitory. A 'relevant connection' is one that connects a neurone representing a recent sensory situation to the neurone representing the action that was taken in response. If a connection between a given concept neurone and action neurone is conducting a signal, then we know that these conditions are being met, so only connections (synapses) that are active need be changed in response to feedback from the environment about the efficacy of the action.

One important problem here is that the feedback might arrive quite some time after the event that precipitated it. For example, if you throw a stone at a policeman the full consequences of your action will probably not unfold until later – much later if you get taken to court. Dealing with this problem of how to propagate reinforcement back in time and relate it to the appropriate cause presents a considerable difficulty to designers of neural networks (such as Ron's brain). The classical solution is a mathematical technique known as the back-propagation algorithm, but for various reasons it is entirely useless in this situation.* Fortunately, we can make the reasonable assumption

---

* The back-propagation algorithm is a mathematical method for working out how much each neurone (in a rather different kind of network) has contributed to any error between the perfect answer and the answer that the network actually supplied, and hence by how much each neurone's influence needs to be altered next time. Back-propagation requires an

that when reward or punishment arrives, it relates to events that have happened relatively recently, and the more recently an event happened, the more likely it is to be causally connected to the arrival of reinforcement. So to solve this problem we can design the network so that when a connection between a concept and an action neurone conducts a signal (representing the link between a situation that is being experienced and an action that is being taken), we make the connection become *susceptible* to reinforcement. We can then let this susceptibility decay slowly over time. When punishment or reward arrives, we can affect the excitatory or inhibitory tendency of every neurone in proportion to how susceptible it presently is. This is not a perfect solution by any means, but it is expedient and will do for our purposes. The problem of reinforcement in neural networks is a live research issue, and much of my work since writing *Creatures* has been concerned with inventing new kinds of network that handle the idea of reinforcement quite differently and do not require it to be propagated back in time like this.

So, a creature finds itself in a particular situation and (for some reason) chooses an action. If the action later turns out to lead to a reward or punishment, we can change the synaptic weights (the amounts by which the synapses modulate the incoming signals) to take account of both the degree and type of the reinforcement and the susceptibility of each synapse. Next time the same situation arises, the creature will have been made either more or less likely to carry out that action once again.

There is a potential snag. Events that just happen to be close in time to a punishment or reward but which are not causally related to it will still get reinforced, and this can lead the creature to the wrong conclusions. Imagine that Ron throws a stone at a policeman and then scratches his nose. Following the inevitable clip round the ear he receives from the policeman, Ron will be unable to discern whether

explicit set of examples, whose correct answers are known, from which to train the network before putting it to use. It also demands that the difference between the 'correct' answer and the one given can be quantified, and that questions and answers follow each other in neat pairs. For a creature living in a complex environment, whether that creature is natural or artificial, there are no right answers – only good enough ones. The degree of error can be hard to quantify, and stimulus and response do not follow each other in tidy sequences. Finally, no prior training is available to an organism – life must be learned by living it. For these reasons (and others, such as the time it takes to perform the calculations), back-propagation is not suitable for our purposes.

the pain was a result of throwing the stone or scratching his nose. But this is often true in real life too – our own brains are pretty good at working out which of the many things we did recently might actually have led to the present circumstances, but sometimes we can link cause and effect only by looking for repeatability. The design of Ron's brain will enable it to do this automatically. If scratching his nose is not causally related to a clip round the ear, then sometimes the one will follow the other and sometimes it will not. Sometimes scratching his nose will be followed (quite fortuitously) by pain and sometimes by pleasure. More frequently it will be followed by neither. On average, the pain and pleasure will balance out, and the connection between 'I am in situation X' and 'I scratched my nose' will be equally punished and rewarded, leading to a net reinforcement of zero. On the other hand, the relationship between throwing stones and being punished will be consistently reinforced.

How to become marginally less stupid over time

This brings up one more necessary bit of finesse. When something bad happens, how much effect should it have on learning? This is a sur-prisingly important question. Imagine that Ron (being 'a Bear of Very Little Brain') puts his hand into a hole in a tree out of sheer curiosity and gets stung by a swarm of angry bees. How much should it hurt? If it hurts too much, Ron will never try that stunt again, which might be a bit unfortunate if other holes contain honey, and honey is what Ron needs to survive. But if it does not hurt enough, Ron is likely to remain in a roughly similar sensory situation to the one that made him choose to put his hand in the hole in the first place, so he may immediately do it again. Of course, if he gets stung several times in a row he will even-tually learn from his mistake, but at the very least this behaviour is not realistically lifelike, and it could even be fatal. New experiences must not be taken as infallible guides to the future, but neither should they be ignored until statistics prove otherwise.

The solution that occurred to me was to make punishment or reward very effective in changing synaptic weights (strongly influencing Ron's future behaviour), but then to allow the neurones gradually to forget the individual experience and remember only a blurred image derived

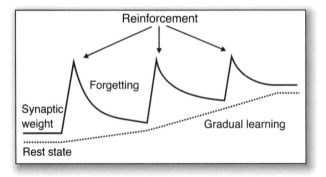

Figure 20. Short-term and long-term memory.

from several cases. This way, Ron doesn't base his entire future on the results of a single trial, which might have been completely spurious. So reinforcement alters the synaptic weight, but over time this weight is allowed to relax back towards a rest state – the synapse forgets the experience. But at the same time I allow the rest state to relax towards the current weight, only much more slowly. When some reward or punishment arrives, the weight will be displaced dramatically from the rest state, and the synapse will learn strongly from the experience. Over time the lesson will be forgotten, but not entirely. By the time the synaptic weight has returned to rest, the rest state will also have changed a little. Over time, repeated rewards will cause the rest state to change, reflecting the *statistical* relationship between the event and reinforcement (Figure 20). So in this way we get the best of both worlds. Whether anything in nature works like this I do not know, but it seems plausible. Recognize the process? Yes, it's our old friend adaptation again.

Kissing oncoming trucks is a bad idea

Earlier I mentioned a second problem that we have to solve. This is the problem of generalization. Given that Ron can perform any of sixteen different actions at any time, he is likely to make a lot of mistakes every time he finds himself in a new situation before he finally hits on the one that works best. In fact it is worse than that. If his first attempt at an action turns out to be fatal, he won't even get a second try.

Real creatures can do much better. When faced with a novel situation, they do not usually try actions at random. Generally, they draw on previous experiences of more or less similar situations. After all, there is no survival value in reflecting that stepping off the cliff was a bad choice while you are plummeting earthwards. It is much better to learn about gravity by falling off your chair a few times, and then noting to yourself how a cliff resembles a very tall chair before deciding what (not) to do when you encounter one. This is why it is so important to let children make painful but safe mistakes. Without such experiences they will have nothing to draw on to help them when they find themselves in a more dangerous situation. In fact, they may not recognize a dangerous situation at all. You may think I'm overstating the case, but we nowadays live in an overprotective society, and children can and do lose life and limb because they fail to recognize what should be quite obvious dangers.

Regardless of how people choose to bring up their own children, I am not going to let this happen to Ron. We need to find a way to bring previous experiences to bear on new and unfamiliar situations, so that the actions Ron tries first are not arbitrary but are likely to be good choices. While I sat on my hilltop contemplating Ron's brain, this issue of how a network can generalize kept nagging at me. I really wanted to find an elegant solution based on self-organizing neural maps. I reasoned that if representations of newly acquired memories either formed close to or migrated towards memories to which they were similar, then all I would have to do is make nerve signals a bit blurry so that they spread out and stimulated nearby neurones. If a novel situation x arose, a memory of x would form near neurones representing similar circumstances, but would not have any action neurones it could connect to, since Ron does not yet know what to do. Yet if the x concept neurone (which is firing, because it represents a situation that is happening) excites its neighbours, then these other concepts will fire too. The influence would grow weaker over distance, and would be proportional to how similar these past memories were to the new one. These other neurones will already be connected to action neurones because they represent situations that have been met and dealt with in the past, so they will begin to excite or inhibit these action neurones. The leaky integration of the action neurones would then combine all the present and recent signals from each of the weakly firing situation neurones, and the network would in effect take a vote on what the creature should do. When a decision has been made and an action is taken, a new connection will form between the concept x and that action. This will later become reinforced if the action that was voted for by the other neurones turns out to have painful or pleasurable consequences. Such a mechanism would work beautifully, and would have the added aesthetic bonus of a brain that organized itself into regions to do with different topics – a sort of family tree of the relationships between concepts.

Sadly I did not manage to fathom a suitable method for such self-organizing behaviour that had all the right characteristics (speed of computation, for example). But it turned out not to matter – quite by accident the mechanism I already had was good enough, even if not as elegant. It all hinges on what 'similar circumstances' means. In a real, highly evolved creature, similarity can be assessed by quite subtle criteria. Poets, for example, use metaphor to bring together ideas that a more prosaic analysis would never have connected. But in the simpler world that Ron inhabits, it is enough to say that situations are similar to one another if they share one or more sensory characteristics. As it happens, we already have concept neurones that represent *sub-concepts* as well as full concepts, and this is enough to solve the generalization problem.

To see how this works, imagine that Ron finds himself being approached by a big green creature on a Tuesday. Sooner or later (big green creatures being such bullies) Ron is going to get hurt. But from this one experience his brain will be unable to tell whether he was hurt because he failed to run away from a big green creature or just from any creature at all, or because creatures are especially unpleasant on Tuesdays, or even because Tuesday is simply not his day.

So that further experiences can teach him to distinguish between these possibilities, I designed Ron's brain in such a way that concept neurones automatically wire themselves up to various permutations of these sensory signals, not just the exact pattern (Figure 21). One neurone will grow to represent the state 'today is Tuesday', and another might connect itself up to represent 'it is approaching'. Neurones that have two inputs will connect to pairs of sensory signals such as 'it is approaching AND it is green', or 'it is Tuesday AND it is green'. For any combination of input signals, neurones will connect up

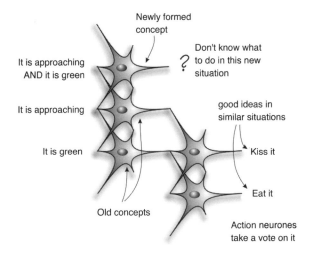

Figure 21. Generalizing from past experience.

to represent most or all permutations of one, two and three sensory inputs (no higher, because our concept neurones never have more than three inputs, unless evolution has intervened and judged that they should be given more). Many of the simpler combinations will actually have been met some time previously – this is probably not the first Tuesday that Ron has witnessed, even though it is the first Tuesday that he has seen an approaching green thing. These situations will therefore already have concept neurones that represent them, including links to any actions that have been taken (and perhaps reinforced) in the past.

So when a novel situation such as 'green AND approaching AND Tuesday' arises, but the neurone or neurones that form to represent it do not know what action they should recommend, there will probably be a whole set of other neurones with useful opinions to offer about similar situations, such as the things that Ron has found rewarding to do on Tuesdays. These sub-concept neurones will then recommend a consensus course of action. The more inputs a concept neurone has, the more signal it receives and so the more strongly it will fire. Consequently, the neurones that share more similarities with the new concept will fire more strongly and have more influence over the vote. All the pros and cons will be weighed up, and eventually a decision neurone will fire. The new concept will form a link with this action and be ready to record the suitability or otherwise of this democratic decision in the specific circumstances. *Voilà!*

This mechanism has some elegant statistical consequences. For instance, the sub-concepts that 'helped out' with recommending an action also learn from any feedback, and this automatically applies in proportion to their degree of generality. There are also some pitfalls, but these are entertainingly lifelike. For example, if the only objects that Ron has previously been approached by are females of his own kind, he may (or may not, depending on how attractive Ron is to his fellows) have concluded that kissing approaching objects is a really good thing to do. This prior experience is likely to backfire on him the first time he meets an oncoming truck. Nevertheless, on balance the mechanism will lead to sensible or at least rational decisions being made in the face of novelty, rather than purely random ones, and will substantially increase Ron's survival chances.

Ron lives (sort of)

This rather technical chain of thought describes the process I went through as I designed Ron's brain. Obviously there are a lot of practical details that I haven't mentioned at all. Despite extensive damping and other ways of keeping the dynamics under control, the design never quite achieved what I had hoped for it. Nevertheless, it did work well enough for the purpose of the game, and I learned a great deal from the experience that I suspect I could never have discovered by any other means.

What we have discussed here is something that I think deserves to be called a brain. It is not remotely like any natural brain I know of, and I am not claiming that it throws any light on how the human brain works at all, except perhaps in the negative sense of demonstrating a few of the drawbacks of simple behaviourist theories of mental processes. It is also pretty stupid in comparison to many animals. But it is a brain, rather than a computer program, and it does, in a rather limited sense, think. The thoughts are not programmed in – the behaviour of the whole structure is not immanent in any of its parts but an emergent consequence of many tiny, concurrent, 'thoughtless' interactions, tuned by nature. Despite being relatively simple in architecture, it demonstrates several characteristics of natural nervous systems, some of which emerged serendipitously, and so I think it has some elegance about it. Armed with a brain like this, Ron can go out into the world a novice and become an expert. He can learn for himself how to survive, without needing a teacher, in a noisy, complex, dangerous environment. Ron is now a bona fide persistent phenomenon.

But a brain is not enough, as we shall see. Now we need to give him a physiology. Time to put away our microscopes and electrodes and get out the chemistry set.

* *

CHAPTER THIRTEEN

THREE PARTS GIN TO
ONE OF VERMOUTH

1 January 1996: Another year; still no product! But RON II LIVES!!!
Finally fixed the major bugs in concept space, decision layer and
instinct genes, and Ron & Eve look much healthier again and a lot
more alive. Language next …

18 March 1996: Gene switching now works, as does crossing-over.
Late-switching genes now overwrite or supplement previous genes
according to context. A birth!!!!!!!!!! Installed chemistry for ovulation
and pregnancy, and temporary scripts for sex and egg-laying. Ron
and Eve gave birth to the first norn bred in captivity!

Excerpts from my programming diary for *Creatures**

Compared with building brains, virtual biochemistry is really easy and
quite fun. In the space of one chapter we shall give Ron the ability to
forget, some pain that he wants to forget, and some pleasures that he
will want to remember, such as eating and mating. To keep everything
logical and allow you to see why all these chemical processes are
related to overall intelligence, we shall start from Ron's brain and work
our way outwards.

Forgetting (if I remember it right)

The first of the two things we need to see to before we can tick the box

---

* These were the good bits – most of the entries relate to endless arguments about the game
design, last-minute changes to the specification, and the amount of time wasted on writing
demonstrators and reports instead of programming; but such is the nature of software development!

marked 'design a brain' is the chemistry for forgetting. As we have seen, the dendrites of concept and action neurones automatically grow towards sources of signal and compete with one another for the right to represent a particular situation or the relationship between a situation and an action. Because Ron is likely to lead a full, rich life, the number of situations he is going to find himself in and the number of actions he will try in response to them are likely to be far greater than the number of neural connections he has available to represent them. Consequently, we need to recycle these connections if they turn out after a while to be unimportant. This is quite easy to do using simulated chemistry. All we need to make is a negative feedback loop that yields a constant supply of loose, uncommitted dendrites – never too many and never too few.

Suppose we design each dendrite so that when it makes a synaptic connection it does so with a given *strength*. We can let that strength fade with time so that the connection will eventually fall loose, releasing the dendrite for reuse. Then, any time a connection shows itself to have value to Ron because it has taken part in an event that led to reinforcement, we can boost its synaptic strength a little so that it takes longer than the others to become disconnected. Synapses will fade and re-strengthen over time, and the more closely any connection is related to reinforcement, the less likely it is to fade out completely. Irrelevant memories will be forgotten, while those that turn out to be very significant (painful or pleasurable) will tend to remain.

Such a mechanism would work fine without any chemical help, except that we have no regulatory control over the process. There is a serious risk that all the connections will be used up, leaving no loose dendrites to accommodate new experiences. Alternatively, they might all fall loose and poor Ron will forget everything he ever learned. But by using a chemical global negative feedback loop we can ensure that a fairly constant pool of uncommitted dendrites remains throughout. If Ron's life turns out to be exceptionally rich and his memories are unusually meaningful, then his forgetfulness will automatically increase in compensation so that he retains the ability to lay down new memories. But if nothing much happens to him, the feedback loop will slow down his rate of forgetting because no more uncommitted dendrites are required, and he will treasure what few memories he has.

The way we do this using chemistry is quite simple, but first I should explain a little about how I defined my virtual chemistry in terms of a computer program, so that you understand what I mean when I start to talk in much more concrete language about 'attaching' chemoreceptors to neurones and 'secreting' chemicals into Ron's bloodstream.

In programming terms, a 'chemical' is simply a memory location that contains a value representing the concentration of a particular substance. A 'chemical reaction' is then a virtual object (a combination of code and data) that 'converts one or more chemicals into one or more products' by lowering the values in some of these memory locations while raising the values in others, to mimic the way that chemical concentrations change during real reactions. Chemoreceptors and chemoemitters (the two 'transducer' building blocks I introduced earlier) are also virtual objects, similar to reactions. But as well as measuring and altering the values of chemical concentrations, they can be 'attached' to other virtual objects in the program, for instance parts of Ron's body or the neurones in his brain. Emitters and receptors provide an interface between the chemistry and the other components of a creature.

When I refer to 'attaching' things, I mean in programming terms that two virtual objects become related by sharing some data. Chemoemitters are virtual objects that are programmed to vary a chemical concentration in proportion to a given value. 'Attaching' a chemoemitter to another object means specifying the source of this value. So, for example, a chemoemitter may be told to compute its chemical secretion based on the output signal from a given neurone. Similarly, a chemoreceptor is 'attached' by specifying a destination for the value it computes as it monitors the concentration of a chemical.

So virtual chemistry is really a series of equations designed to mimic the complex dynamics of a container filled with chemicals that react with or catalyse one another. The container is Ron's 'bloodstream', and its contents are monitored by chemoreceptors, which supply data to other virtual objects. At the same time, chemoemitters monitor the data in other objects and increase the concentrations of certain chemicals as necessary. Ron's bloodstream is therefore a kind of computer (reminiscent of the old analogue computers that preceded the modern digital form) whose inputs and outputs are connected via receptors and emitters to his brain and body. I shall make no further mention of equations and memory locations, and shall drop the quotation marks from metaphors like 'attaching' and 'reaction'. After all, I am trying to persuade you that Ron is a real, living creature and not a metaphor for one or a simulacrum. The components are not real but Ron will be, and it is much better if we treat these chemicals, receptors and emitters as if they were objects that can be plugged together, rather than computer code. Our job is no longer that of a programmer, but of a bioengineer.

So, to return to the first problem to be solved using our chemical computer. We want to create a system of negative feedback that regulates the breakdown of connections between neurones and ensures that there is always spare capacity for laying down new memories. To do this, we attach a chemoemitter to each dendrite in such a way that the emitter produces a certain chemical whenever the dendrite is unconnected. The chemical is designed to decay with time, so the concentration does not build up indefinitely but finds an equilibrium level (this is a leaky integrator again). The overall concentration of the chemical is therefore proportional to the number of loose dendrites in the brain at any moment. All we then have to do is attach a chemoreceptor, sensitive to the same chemical, to the property of a neurone that controls the fading of synaptic strength. The result is a simple negative feedback loop: the more loose dendrites there are, the more the chemical will be produced and the more slowly the synaptic strength of the remaining dendrites will fade, eventually preventing any more connections from coming loose. As new connections get made, the pool of loose dendrites gets smaller, the concentration of the chemical goes down and the rate of atrophy increases, bringing more dendrites into the pool again.

This feedback loop, even though it is specific to a particular class of neurone (there is one such loop for the concept cells and another for the action cells), is *global*. The chemoreceptors and chemoemitters transduce a flow of information from directed to diffuse form and back again. Local information about individual dendrites contributes to the global information shared by them all, which then regulates local changes to individual atrophy rates. This is rather more 'democratic' and bottom-up in feel than it would be if some 'control program' went around counting loose dendrites and then decided which connections were for the chop – the difference is subtle, but it is there.

This mechanism may seem rather overblown – why all the fuss with chemicals and stuff; why not just code the behaviour straight into the neurones? It does have its advantages, however. One has to do with the fact that the emitters, reactions and receptors are coded genetically. Relapsing briefly into computer-speak, this means that there is a long list of numbers that the program uses to construct a creature. These numbers define the parameters and interconnections of all the virtual objects – receptors, reactions, neurones, and so on – that comprise the creature. The structure of a creature is therefore defined not by immutable program code, but by a list of easily modified numbers, and this list gets passed down from parent to child. In other words, although we have designed this particular chemical network to solve a problem in the way we think is best, we are also allowing evolution to tune the system in successive generations – perhaps natural selection will discover a more optimal balance of loose dendrites, or even change the mechanism completely.

Another advantage is that bringing the feedback loop into the chemical domain allows us to add new features to it or make use of information encoded in it as freely as we wish. In this particular case we have no further use for the mechanism, but later we shall tap into chemistry that was developed for one purpose and use it for others. All this extra functionality comes for free – playing with Lego bricks is much easier and more intuitive than writing computer code.

Driven to despair

We have solved the problem of forgetting, but we need another bit of brain chemistry to handle reinforcement. We need to punish and reward the creature by changing certain synaptic weights so that it can learn from its mistakes. But what exactly is reinforcement, anyway? So far I have talked about reward and punishment as some sort of nebulous pain or pleasure. Pain at least is a real sensation, and any response to a situation that leads to physical pain should be punished so that the response becomes less likely the next time the same situation arises. But not all reinforcement comes from pain. What else causes reinforcement, and how do we measure it?

One way to tackle this problem is to imagine that the creature has a wide variety of 'drives' or 'needs', and arrange for changes in these drive levels to produce the necessary punishment or reward. Take hunger, for example. Another way to describe hunger is 'the drive to eat'. A newborn creature does not realize that it needs to eat when it becomes hungry, and so it simply gets hungrier and hungrier. But when the creature does something that reduces hunger (say it randomly chooses to put something edible in its mouth) we want to reward it for taking this action. So reductions in drives should emit a rewarding reinforcement. Similarly, an action that increases a drive (such as loneliness, perhaps) should generate a punishment signal to the brain.

So suppose we define a series of drives and represent them using chemicals. One chemical represents the amount of pain the creature currently feels, another the level of hunger, and so on. We can define as many drives as we wish: loneliness, overcrowding, tiredness, anger, fear, hotness, boredom, and so on. The more of each chemical there is in Ron's bloodstream, the hungrier, hotter, angrier or whatever he must be.

To provide the reinforcement we need to set up some chemical reactions such that any activity which increases the level of a drive will create a 'punishment' chemical, and anything which decreases a drive will emit a 'reward' chemical. By attaching chemoreceptors to the appropriate synapses we can then change the synaptic weights (and synaptic strengths too, to decrease 'forgetfulness') of any connections that are currently susceptible to reinforcement, according to how much Ron's drive levels have changed (Figure 22).

It is all very well rewarding Ron for eating when he is hungry, but that is not all we have to do. Suppose that eating when he is already full is a bad thing to do, as it is in nature (why expend energy on eating when you are fully replenished?). In this case we should administer a punishment. But how can Ron tell the two cases apart? Sometimes he eats and is rewarded; other times he eats and is punished. Unless he knows how hungry he is at the time, he cannot distinguish the first case from the second. So it is vitally important for Ron to be made aware of the present intensity of each drive, as well as the reinforcement that will follow changes in it. This way, his brain can build up memories such as 'if it is food and I am hungry then I should eat' and 'if it is food [and I am not hungry] then I should not eat'. His behaviour will then be controlled by the degree of hunger he is feeling when he

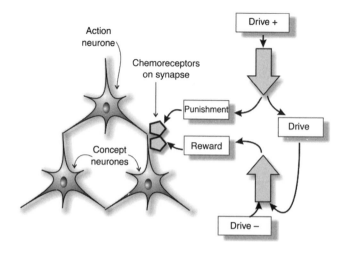

Figure 22. Part of the reinforcement chemistry.

sees food. The same applies to anger, sex drive, and all the others. Adding this ability requires no programming – we can simply attach chemoreceptors to a subset of the neurones that feed Ron's brain with sensory information. Each receptor can be sensitized to a particular drive chemical, and the signal emitted by the neurone will then increase or decrease in response to the level of the drive.

This network of reactions is a surprisingly flexible and potent system. The list of chemical drive levels gives Ron a 'bar chart' of signals describing his fitness for survival, because he can tell how much and in what ways his present state differs from an ideal state in which all drives are zero. His brain is aware of this information, and the reinforcement mechanism allows him to relate changes in drives to the behaviours that caused them. Inside his brain's learning mechanism the reinforcement consists only of punishment and reward, but in his bloodstream there is a wider range of ways to modulate and express it.

The overall effect can be subtle and very powerful. For example, many events in the outside world (or in Ron's own physiology) might alter more than one drive at a time, and this can have useful effects. Perhaps Ron is bored (we specify boredom as a drive that increases steadily with time), and he chooses to play with a ball. The act of bouncing the ball might change several drive levels at once – perhaps it decreases his boredom substantially but slightly increases his tiredness and hotness. If Ron was very bored when he started to play with the ball, the net result of these changes will be a reward because boredom is being reduced much more quickly than tiredness and hotness are increasing. But once Ron has ceased to be bored, the level of this drive chemical cannot be reduced any further. So the net result of the stimulus is now a punishment, since only tiredness and hotness are being affected and both are increasing. Because Ron is aware of the state of his drives, he learns to distinguish between good times and bad times to play soccer, and he will not just go on playing for ever.

A creature marches on its stomach

So we have a wide variety of ways to modulate learning, and Ron can be aware of a number of bodily states that give him the information he needs to survive. Some of these drives can be changed quite simply and explicitly by external stimuli: for instance, pain levels increase directly as a result of physical traumas such as walking into a wall or trying to stick one's hand in a bees' nest. Other drives, such as hunger or loneliness, need to be modulated in more sophisticated ways. The necessary information cannot be obtained directly, so we must develop some systems that perform more subtle chemical computations to tell us what we need to know.

We want Ron to have to learn to fend for himself, since the survival urge is what drives his intelligence. One of the simplest ways to ensure this is to give him an internal energy store that is used up by activity and replenished by eating. If Ron fails to maintain this energy store, we must ensure that he dies. He will therefore have to learn what to eat, learn to balance eating with other activities and learn ways to avoid spending more energy on finding food than the food provides in return. In other words, to give Ron something worth thinking about we should give him a digestive and respiratory system. To do this we need a chemical to represent Ron's internal energy resource. We can give it a biochemically appropriate name – glucose. When Ron eats, we need to increase his glucose levels; when he expends energy, we need to decrease them. If Ron runs out of glucose, he must die.

In fact, glucose levels are likely to fluctuate quite rapidly, and we don't want Ron to die the very first time he learns the meaning of hunger. So it would help if we also had a longer-term energy store, which builds up in times of excess but can be converted back into glucose when the need arises. Most natural organisms have at least one such long-term storage system. Humans can draw on several alternative sources of energy when necessary, including our own muscles in times of serious malnutrition. For Ron, a single long-term energy store will suffice, and again we can give this a plausible biochemical name – glycogen.

To convert excess glucose to glycogen during times of plenty and back again during times of starvation, we need a reversible chemical reaction. We can achieve this by defining two simple reactions back-to-back:

glucose → glycogen,   glycogen → glucose

By varying the reaction rates we can ensure that glycogen builds up slowly when there is more glucose present than glycogen, but turns back again when necessary to prevent the glucose concentration from hitting rock bottom. A chemoreceptor can then be used to trigger the creature's death if glucose concentration falls to zero, which will happen only when all the glycogen has been used up.

When Ron eats, we can program the virtual food objects to supply him with one or more chemicals (starch and sugar, say), each of which gets slowly converted into a characteristic amount of glucose in his bloodstream. This is digestion. Finally, we need to burn glucose at a rate proportional to Ron's physical activity level. This is respiration (respiration is the oxidation of glucose to provide energy, not the act of breathing, as people sometimes mistakenly think). When I reached this point while writing *Creatures,* I found, rather conveniently, that I already had a location in the computer's memory in which I stored the number of limb movements that had occurred during the most recent tick of the time-slicing clock. All I had to do was attach a chemoemitter to this memory location and set it to emit an 'enzyme' at a rate determined by the number it found there. This enzyme was then made to react with the glucose, removing it in proportion to the amount by which Ron moved his body. The result of the reaction was the production of carbon dioxide and water, as in real life. Sadly, the marketing department prevented me from providing a suitably biological exit route for the waste water.

There is only one trick left to perform: somehow we need to modulate Ron's hunger levels in response to his feeding behaviour so that we can provide reinforcement signals to control his learning. The best way I found to do this was to produce hunger-increaser and hunger-reducer chemicals as by-products of the two-way reaction between glucose and glycogen. When glucose is being converted to glycogen, we can infer that Ron has eaten recently and all is well. On the other hand, when glycogen is being converted to glucose it must be because glucose levels have fallen, so we should increase Ron's hunger level.

So, with a handful of chemical reactions and a few strategically placed chemoemitters and chemoreceptors we have built Ron a digestive and respiratory system (Figure 23). Every step in this system has some real purpose – none of it is there just for show.

No more Mr Nice Guy – Aaaaaachooo!

The same cannot be said for the next chemical system. This one is pretty gratuitous, but I am bored with being a beneficent digital god. It

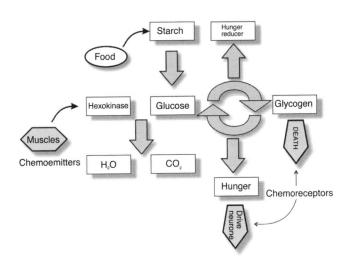

Figure 23. Digestion and respiration.

is time for a bit of digital devilment. Suppose we add some simulated bacteria to Ron's world, and specify that each bacterium is capable of emitting a few chemicals into the bloodstream of any creature it infects. Some of these chemicals may be toxic, interfering with the natural processes of respiration or whatever. We could even define a chemical that we shall call histamine, and through the magic of chemoreceptors we can cause histamine to trigger an involuntary reflex from the creature: a sneeze or a cough. Furthermore, we can contrive it so that bacteria are passed from one creature to another through physical contact or via a histamine-induced sneeze. Coughs and sneezes spread diseases! Finally, we can design the bacteria so that the chemicals they secrete are encoded in their genes and capable of mutating. Consequently, the bacteria that are passed from host to host most effectively will increase more quickly in the population, and our diseases can evolve to become more virulent.

In compensation for this new stress in Ron's life, we can do two things. For a start, if we can have toxins (whether emitted by bacteria or by other objects in the environment, such as poisonous fungi) then we must be able to create medicines too. By defining chemical reactions that interfere with various aspects of the biological processes, toxins can disrupt respiration, digestion, reproduction, memory regulation or learning. But other chemicals can be used to combat these toxins, either by reacting with them to remove them from the system, by blocking them or by supplementing the processes that they have destroyed. The world is our oyster in this respect – we can make up toxins and medicines to our hearts' content, since there are so many points in the network of reactions at which we can interfere with the system. We can administer the medicines deliberately or we can let the creatures discover them for themselves. In the *Creatures* game I placed many active chemicals into 'herbs' that grow naturally in the virtual environment, for the creatures and their owners to discover and learn to use.

The other thing we can do is give our creatures their own natural defence against infection – an immune system of sorts. Our own immune system recognizes the presence of invaders by detecting the existence of characteristic substances on their surfaces called antigens. The antigens then stimulate the production of highly specific proteins called antibodies and begin a cascade of activity that results in the destruction of the infecting agent. Our immune system is tremendously complex, but we can simulate the essence of it in our creatures more simply.

We can 'coat' the virtual bacteria in a random assortment of chemicals that we can call antigens, and define an equivalent list of chemicals and associated chemical reactions in the creatures to represent antibodies. We can then arrange for the chance of a bacterial infection taking hold to depend on whether the creature has sufficient antibodies that match the antigens on the bacteria. We can give every creature a smattering of antibodies from birth, but define their chemistry so that each antibody will build up naturally in response to infection by bacteria that have the complementary antigen. This way, Ron will develop a gradual immunity to disease, following childhood infections, and become capable of fighting off future insurgents before they are able to take hold.

Of course, we cannot let the creatures have it too easy – the antibodies should decay over time, and we can also inhibit their functions or prevent their production in response to environmental stress or old age. We can create the equivalent of stress by emitting a rapidly decaying chemical (adrenaline is the nearest biological equivalent) whenever certain drives exceed dangerous levels. This adrenaline can be converted into another longer-lasting chemical that builds up slowly over time and interferes with other reactions to produce the harmful effects of stress.

Not long after *Creatures* was published, two German medical students sent me a charming document. It was a full scientific paper, describing one of these disease states and outlining its diagnosis, prognosis and treatment in proper medical language. They therefore have the honour of naming and describing perhaps the world's first virtual disease (if you don't count computer viruses): Schrey–Leonard syndrome.

The wrong time of the hour

Back to less frivolous biochemical matters – this time the reproductive system. The primary purpose of this system is to enable creatures to pass their genes on to another generation, but we might as well make the process as subtle and interesting as we can. No other activity in nature involves such behavioural sophistication as sex. From the peacock's tail feathers to the social nuances of the nightclub, the process of finding, attracting and getting it together with a mate is an advanced and absolutely vital activity. Let us see how much of this complexity we can build into our virtual life forms.

We need two sexes, and they should be different. Males should donate sperm (their genetic code) and females should combine this with their own genes and carry the resultant child. The male reproductive system that I designed is suitably straightforward: males have a sex drive chemical, which starts to increase in concentration around the age of puberty and then just goes on rising and rising ... Sex can reduce the drive substantially, but in no time at all it will be back to its old level.

Females are rather subtler, on the whole, and I especially wanted to make my female creatures (the first of which was called Eve, by the way) have some kind of hormone cycle. For this I needed an oscillator, and I soon discovered that I could produce one by attaching a chemoemitter back-to-back with a chemoreceptor. The emitter produces a hormone (we shall call it oestrogen), the concentration of which is measured by the receptor. This receptor is configured in such a way that when the concentration of oestrogen reaches a critical threshold, the receptor switches off. This turns off the emitter, and so no more oestrogen is produced. That which is left in the bloodstream decays gradually until the concentration has fallen low enough to flip the chemoreceptor back into its 'on' state, whereupon the sequence begins again. The whole cycle takes around an hour or less, rather than the more familiar month.

Using another chemoreceptor attached to a particular location in the computer's memory, I could control the creature's fertility level so that she was fertile only during the falling half of her ovulatory cycle (Figure 24). If she was injected with a male genome (sperm) during this period, she would perhaps conceive, and this would trigger another chemoemitter with two consequences. First, another hormone (which I dubbed gonadotrophin, by analogy with the nearest biological counterpart I know) starts to rise rapidly. This has the effect of suppressing oestrogen production and cutting off the female's ovulation cycle. The other, progesterone, rises much more slowly and

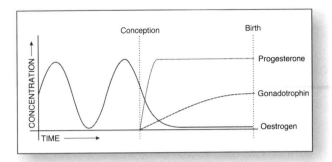

Figure 24. A virtual pregnancy.

marks the progress of pregnancy. When it reaches a high enough concentration, it triggers the chemoreceptor that results in a birth (the
creatures lay eggs, incidentally).

These hormones are also able to affect various other chemicals. For
instance, the female's sex drive is controlled by oestrogen, which
makes her more interested in the opposite sex and more likely to conceive during the critical part of her cycle. One of the other things that I
made capable of affecting the female sex drive was a pheromone produced by males – a sort of virtual musky scent. I made them produce
this substance whenever they chose to perform a specific behaviour in
the presence of females – a little mating dance, or lek, to be precise. The
idea was that males who learn to do the dance should successfully
impregnate more females than those who do not, so their genes ought
to increase in the population. A genetic predisposition to perform the
dance should therefore become more apparent over many generations. But this dance is rather subtle, and its effects are hard to discern.
To this day I don't know whether the mating ritual has ever been
witnessed and recorded by any of the people who keep and study
Creatures.

Like father, like son

Because the whole recipe for creating a creature (the list of configuration numbers) is wrapped up in its genes, and the male genes are

passed to the female during sex, reproduction involves more than just mating behaviour. All the genes that go to make a creature (some three hundred in the initial generation) are arranged along a single long 'chromosome'. During conception, the male and female chromosomes are intertwined and broken at the points where they cross. From the fragments, two new chromosomes are derived – each containing roughly half the genes from mum and half from dad. These chromosomes are called haploid, because there is only one copy of each gene in a single creature. Our own chromosomes come in pairs and are called diploid. The diploid reproductive process results in a fertilized egg containing one full set of chromosomes, exactly half the genes coming from each parent. My creatures are slightly different, and every conception ought to result in twins because mum's and dad's single chromosomes get split and recombined to make two new chromosomes – enough for two children. To get round this, I simply destroy one of the child chromosomes, leaving the other to be used as the recipe for making a baby.

So a baby creature will have a random selection of genes from each parent. Its appearance, movement, brain structure and biochemistry will therefore be similar to but subtly different from either parent's. There are also two other ways in which the infant's chromosome can differ from those of its mother and father. The first of these is mutation: occasionally, one or more of the genetic numbers is altered at random. Since each of these values represents a parameter or a connection rule for a group of neurones, a chemoreceptor or whatever, the random change leads to the production of a component in the offspring that behaves, or is connected, differently from the equivalent in the parent. For instance, a chemoreceptor may start to respond to a different chemical, or it might find itself attached to a different virtual object entirely. Mutations can obviously produce both subtle and profound changes.

The second type of error during conception is gene duplication or omission. A whole gene (a gene is the list of numbers that define a single virtual object) or even a string of them might get left out of the new chromosome entirely, and the corresponding structures will fail to grow in the resultant embryo. I once had a creature sent to me from Australia, by email, from its distraught owners. They told me it had simply sat motionless from birth and was now in a state of extreme starvation, so would I cure it for them please? I got out my genetic manipulation tools and looked closely at its genes. It turned out that a crucial gene had become deleted, leading to the non-appearance of a complete cluster of sensory neurones. The poor creature was consequently deaf and blind and had no idea that there was a world out there to explore, which is why she had just sat there and starved. After much ado, I managed to reinsert the missing gene and cause the crucial brain structure to grow. I then fed the creature up, taught her a few swear words and mailed her back to Australia. That Christmas I received a card from the family, to tell me that little Kelly was now doing just fine.

Instead of a gene deletion, it sometimes happens that a gene becomes duplicated. The consequences of this depend on the nature of the gene, but in general it means that there are now two structures where before there was only one. Sometimes this extra structure will have an effect (such as doubling the output of a chemoemitter), and sometimes it will have no impact whatsoever. Nevertheless, because it does not represent a critical part of the creature's make-up, it is free to mutate in future generations without harming the organism. If it eventually mutates into a form where it gives rise to a new structure that confers some advantage, the creature can be said to have evolved creatively: instead of just tweaking and tuning what it already had, the genome has discovered a new feature all by itself. The creatures therefore have the potential, over a long period, of evolving novel structures that I hadn't thought of when I designed them. The limits to what might emerge are really quite loose, and so there is no telling what might turn up, given enough time. There are certainly creatures around with more than the original number of neurone clusters in their brains, and I can think of ways that further clusters might be useful, but I have not seen any evidence to date that these extra neurones perform any function.

So far, nobody has carried out a proper scientific study of the *Creatures* gene pool, so I do not know what, if anything, has happened to the species. Perhaps very little – after all, evolution takes place only over a large number of generations, and these creatures take some time to live out their little lives before they either find a mate and reproduce, or else die from starvation or disease. Their evolution rate may not be very many orders of magnitude faster than our own, despite their comparative simplicity. Nevertheless, the world population of these creatures is quite astoundingly large – perhaps several million are in existence at any one time (for comparison there are a little over a million elephants in the world). And the more enthusiastic owners swap creatures and deliberately cross-breed them with all the fervour of dog breeders or pigeon fanciers, so who knows where they might end up?

Further duties of a creator

There is so much more to tell, but too little space to tell it. We haven't talked about ageing, for example. By judicious use of chemical decay and thresholds on chemoreceptors we can cause the creature's genes to be re-read at intervals. Each gene is given a certain switch-on time, and so each time the list is read, new genes may turn on and bring new chemistry into play, such as the growth of the reproductive system during puberty. As a creature starts to get old we can, if we so choose, make it more infirm, for example give it slower walking gaits, poorer conversion of food into energy or a weakening immune system. Finally, in principle, we can trigger the death of a creature through old age, to make room for the next generation. This is what I intended to do in *Creatures*, but I realized some time after the game had been published that I had missed out a crucial gene. So the fact that everybody's creatures had been dying of old age had nothing to do with this deliberate euthanasia at all. They were dying because they had become more susceptible to disease, or could no longer reach sources of food quickly enough. It was quite satisfying to find out that they were all dying from natural causes.

We haven't talked about instincts, either. They are too complicated to justify another long explanation. Suffice it to say that a fully unformed brain at birth is useless. Although boredom and other basic drives will eventually provide the feedback that drives behaviour, the creature will initially have no incentive to move at all, and so will never taste the experiences from which it could learn new behaviour. To counter this we must genetically wire up the neonate's brain with a few basic instinctive or reflexive behaviours, such as 'if you are bored, standing still is not a good idea'. Unfortunately, because we have relinquished control of the creatures' brains to the machinations of evolution, we have absolutely no idea how they might work after a few hundred generations of natural selection have had their effect. So we cannot define genes that explicitly wire up individual neurones to provide these instinctive reactions because we have no guarantee that future brains will operate in the way we expect. Luckily I dreamed up a solution that allowed baby creatures to *learn* instincts while still inside the egg, with genes providing the stimuli and reinforcement in much the same way that the environment would do after birth. This was a neat solution because it was independent of the learning mechanism itself – as long as the creatures' brains still learn to relate stimulus to action through reinforcement, regardless of the means by which they do it, the instinct genes will work.

Incidentally, with the right selection of instincts to interact with the creature's drives we can emulate some quite subtle lifelike characteristics that at first glance seem to make no sense in terms of simple chemistry and nerve signals, such as curiosity or various aspects of social behaviour. Curiosity, for example, is a striking characteristic of many of the more intelligent species – cats, birds, chimps and humans. It provides us with many of the experiences that enable us to learn. We can create something fairly similar in our creatures by combining an instinctive restlessness in the face of an unchanging environment with an improvement to the attention mechanism that makes new objects seem more 'interesting' than familiar ones. When nothing much is happening the creatures will tend to go off and explore, and they will not be able to resist the temptation to fiddle with newly discovered things.

Finally, we have not talked about language. This is an area I wish I had had more time to explore properly while I was writing *Creatures*. In the event, I had to resort to conventional machine learning techniques to allow creatures to learn new words and their associations. They can pick these words up by hearing their owners 'say' them (for example, draw their attention to a ball and then type the word 'ball', and they will begin to associate the one with the other). Alternatively they can 'read' them from labels and signs in the virtual world or from the screen of a virtual computer that is running an 'educational software package'. The only time their brains play a part is when people ask them to do something. The sentences that their owners type into the program are scanned for known verbs (words they have learned to relate with actions they understand) and nouns (words that relate to the objects they know). Words of each type are then presented to the brain along different sets of nerve fibres. So nouns are fed into the attention mechanism, where they may or may not persuade the creature to pay attention to an object of that type, while verbs are fed into the concept neurones, which have previously been wired up by instinct genes to connect them with their associated actions. So if you say 'press the button' to Ron, his attention may be drawn to any button that he is currently aware of, and if it is, then the verb will encourage him to press it. Of course, because this command passes through the neural network, the instinctive associations can be un-learned or modified by experience, so if you try to make your creatures do unpleasant things, they'll soon learn to ignore you.

This is not real language because there are no rules of syntax, and the kinds of things you can say to your creatures are very limited indeed. The things they can say back to you are even more restricted – they occasionally combine the names they have learned for actions and objects to 'tell' you what they are doing, and sometimes use 'analogies' to explain that they have a pressing need (such as 'Ron run' or whatever words they know for those things, meaning 'I am scared'). In neither case do the creatures really choose to speak, nor can they learn new ways of expressing themselves. Nevertheless, the creatures' proto-language is sufficiently important in their day-to-day activity that it once prompted a discussion on the Internet concerning which language people should teach their creatures so that they would not feel uncomfortable when they were transferred across the Net to virtual worlds in other countries ...

Proof of the pudding

So there we have it. By combining simple cybernetic building blocks such as modulators, leaky integrators and oscillators into massively parallel, directed and diffuse feedback networks, we can make something that is not only alive in the technical sense but also alive in the richer, more rounded sense too. These creatures are not very smart, but they do have individual little personalities. They live out independent lives and behave in ways that I, as their creator, didn't program them to and sometimes didn't even expect. They also elicit appropriate responses from their owners. I shall tell you a little more of this in the next chapter, but for now I will leave you with just one of the thousands of pages that people have published on the Internet about their creatures' exploits. It is surprising what a little bundle of neurones and simulated biochemicals can get up to! The author of the following Web page, Lis Morris, created a new breed of creature by the deliberate and skilful genetic manipulation of a naturally occurring mutant variety. She named her creatures Hippy Norns because they have been genetically modified to turn glucose into happiness (the equivalent of a chocolate fixation, perhaps?) and here she explains how to look after them.

The Hippy Norns are a genetically engineered breed of norns, with a very altered biochemistry. They utilise energy to stay happy at all times! I called them Hippy norns, because, as you will soon find out, all they need is love!

However, they can be difficult to look after, and require some expertise as a Creatures user. [Read on] to find out more about looking after your hippy norns.

So how exactly are they different? I altered many of their biochemical reaction genes, and added 17 new genes. Lot of work, huh? These norns are so different to normal norns that I thought I'd better make up a file to tell you some of the basics of looking after them, without going into technical detail.

Starting at the beginning – how do I look after baby hippies?
These norns as babies are no different to other norns – except that teaching them is a little different. They get rewarded for doing something that reduces a drive (e.g. need for pleasure, anger, etc.) whether they need it reduced or not. This means that they 'fiddle' with things a lot … always picking up items and using them.

One major difference is that you don't have to teach them to feed. A tickle on the head, or playing with a toy is the equivalent of a good meal. However, it is a good idea to get them eating, because it can boost their health levels. A very newborn hippy norn is often very slow to get moving. They are not, however, 'children of the mind' mutants. I think they just feel so blessed out they feel no need to move.

Is there anything that is really dangerous to them?
Yes definitely! Anything that causes a norn a lot of pain – grendels or

the beehive update – can kill a hippy norn. Though they will not be in pain for long, controlling the pain takes a lot of energy. Boredom also seems to be very bad for them, too, though this tends to equilibrate with time, the norn settling down to a lower health level. It is a general rule that anything that would make a normal norn unhappy will reduce a hippy norn's health level. However, a quick tickle on the head is the best medicine for anything!

Can a hippy norn get sick like a normal one?
Yes, they can catch diseases of normal norns, but they have a fully functional immune system. If their health starts to drop, then try tickling them to increase their health levels. If this does not work, try slapping them, waiting a few seconds, then tickling them. It may sound odd but it works! If their health decreases greatly, treat them as you would your normal norns.

What's happening? This norn's health has dropped to 40%! What can I do?
Check the norn in question for obvious sickness. If there is nothing wrong, then the norn is just fighting off boredom/anger/something else. If you look at the drive levels in the science kit, you'll probably see one that is oscillating low down. This is the drive that the norn is using its energy to control. A common one is boredom. Try to get the norns into a group of other norns, or give it a toy to play with. You could even simply give it something to eat – this raises health levels just like in normal norns. I have one norn, Prime, who always has fairly low health – around 20–40%. At first this panicked me, but then I realised that she slaps *constantly* and gets slapped back! Serves her right …

Can I breed these norns with my normal ones?
Well I've found that these norns … breed! Like Rabbits. My original eight hippies – four genetically engineered plus four second generation ones, have produced 17 eggs! The females spend all their time pregnant or recovering. As for their ability to breed with normal norns …
urm … pass! I haven't tried it yet, but I think that if you do you'll get either an immortal norn, or one that will die very quickly, or possibly both. You'll get a mishmash of normal and hippy genes … anything could happen! The definition of a new species in biology is a group of animals able to breed with any animal in that group, but not to produce successful offspring with any other animals. Does this make hippy norns a new species?

And on that note, I'll leave you to go and play with your new norns!
Web page by Lis Morris (http://www.shee.demon.co.uk/health.html)

CHAPTER FOURTEEN

* *

TAKING OVER THE WORLD

The human race as we know it is very likely in its end game; our period of dominance on Earth is about to be terminated. We can try and reason and bargain with the machines which take over, but why should they listen when they are far more intelligent than we are? All we should expect is that we humans are treated by the machines in the same way that we now treat other animals, as slave workers, energy producers or curiosities in zoos.

Kevin Warwick, *In the Mind of the Machine*

A short while ago I gave a talk at the Institute for Contemporary Arts in London. My co-lecturers that night were Johanna Moore and Mark Humphrey, both of whom are professional AI scientists. To be honest, I think all three of us found our talks rather hard going. I have given quite a few lectures to people from all walks of life, and usually they laugh at my jokes and show genuine enthusiasm for the subject. But this audience only chuckled politely and seemed a little reserved throughout. It was only some time later, after having fended off questions from the floor about strange topics such as capitalist conspiracies, that I realized what the problem was. Most of the lectures I have given have been to people who were at the very least interested in science and positive about it. But these people were primarily from the arts, and for some reason had quite a different mindset. To them, artificial intelligence was something to be scared of or angry about, or even to deny the very possibility of. They seemed to lump the idea of intelligent living machines alongside such fear-inducing things as genetically modified foods and nuclear power.

Perhaps it didn't help that the evening of talks had taken its name from one of the section headings of this book, 'Whatever happened to HAL?'

Perhaps they had been expecting us to tell them how frighteningly close we were to creating human-level artificial intelligence. Maybe they expected demonstrations of the cold, calculating, ruthless, mechanical warriors that we were almost ready to unleash on humanity. It did not seem to help that my own talk was on why fifty years of AI had been such a dismal failure. But actually I do not think the muted reception was our fault. I think this audience had been exposed to the dire predictions of too many doomsayers. Some of these false prophets are actually AI researchers themselves, yet they see nothing wrong with painting a terrible future for humankind in which machines rapidly surpass us in intelligence and we become little more than a source of slaves or domestic pets for the amusement of the master machines. *But this is not how it is going to be at all!* Machines are not going to take over the world. Artificial intelligence will not create monsters. Almost the opposite, in fact. Perhaps we should look at some of the 'logic' that causes people to be so fearful.

IQ-phobia

In the first place, there is the assumption that intelligence in itself is something to fear. Personally, I do not find intelligent things frightening in the least – it is the stupid ones that scare me. The more intelligent a person is, the more likely their behaviour is to be rational, thoughtful and considerate. I am sure that this will apply equally well to machines. We are more ruled by machines now than we ever will be in the future, not because machines are smart but because they are stupid. We are slaves to our word processors and databases because they do not understand what we want of them and are not smart enough to work things out for themselves, so we have to adjust our own behaviour to suit them.

I accept that many of the tyrants of the past have been intelligent people – you simply do not get to be a dictator unless you are smart, so it is almost inevitable that Hitler, Mussolini, Stalin and Genghis Khan were intellectually capable people. Not intelligent enough, perhaps, to overcome their innate antisocial, megalomaniac, pugnacious character flaws, but these traits are not confined to intelligent people. Most malicious leaders are smart, but it does not follow that most smart people are malicious.

A related assumption is that intelligence equates to heartlessness and ruthlessness. In fiction, superintelligent beings are often portrayed as emotionless creatures who stick to their logic regardless of how callous it may be. Actually this has more to do with poetic licence and perhaps the poor logic of the authors than with the supposedly impeccable reasoning of the hyper-intellects they write about. There is nothing fundamentally wrong with the process of logical reasoning, as long as one takes into account *all* the relevant factors, including people's emotions and the effects an action has when looked at from other people's viewpoints. If the superminds were really so smart, they would take all this into account and not make such stupid and calamitous decisions.

But there are other reasons for the myth that emotion and intelligence are mutually exclusive. *Star Trek*'s Captain Kirk and Mr Spock represent the two sides of this supposed dichotomy: Spock is ruthlessly logical while Kirk, although bright, is driven by his heart more than his head. Kirk, of course, always comes out on top. Partly I think this characterization stems from observations of human nature that have been amplified out of proportion. Certainly there are some people like Spock, for example the musical, mathematical or artistic idiot savants whose unusual skills are related to autism. They certainly tend to lack any sensitivity to social cues or other people's feelings, while at the same time having remarkable mental talents. But no general conclusions can be drawn from these rare and unfortunate people.

Partly the dichotomy comes from AI itself, for many researchers have believed that intelligence can be abstracted out and programmed straight into a computer. They have assumed that there is no need for 'base animal urges' such as emotional states in such systems, and intelligence can come from logical reasoning power alone. They are just plain wrong. Fifty years of abortive AI have shown their error. I hope this book will have done a little to redress the balance: emotions are an important component of artificial intelligence, just as they are of natural intelligence. Intelligence is, furthermore, a much richer phenomenon than reasoning alone. Anyone who thinks they can make a machine with the general intelligence of a human being but no emotional or commonsensical qualities at all is wasting their time.

There is some truth to the argument that machines will want different things from humans, and at some point there may be potential for conflict because of this. But this argument is based on a shallow view of how the future will pan out. After all, it is we who shall make these machines; we who shall define for them what their drives and needs are, and what they find enjoyable and desirable in their little metal lives. Why would we design machines whose needs conflict with our own? If we build intelligent cars, they will be designed to take pride in their ability to carry us to our destinations quickly, efficiently and without knocking over a single pedestrian. They will enjoy a good drive as much as we enjoy a good night out. Why on earth would we design them any other way? And just because we give them a degree of autonomy, it does not mean that they will immediately take up a new, conflicting set of desires. They will be just as trapped inside their own drive and reinforcement regime as we are in ours. We humans rebel against our circumstances precisely *because* we were not designed to enjoy much of what in modern society we spend most of our time doing.

One of the underlying assumptions from which many of these other arguments spring is the belief that machine intelligence will run away with itself: machines will get smarter quicker, and will rapidly surpass us in intelligence. This myth has several origins. One is the same fallacious argument mentioned above: that intelligence can be abstracted from its substrate and somehow 'streamlined'. From this it is deduced that a logical reasoning machine like a computer will be so little weighed down by the biological baggage that supposedly cripples our own minds that it can race ahead and become rapidly more intelligent. This argument fails on many counts. First, intelligence cannot be abstracted – a machine that is as intelligent as a human would in all probability share most of our mental attributes, not just 'pure reasoning'. Second, most so-called reasoning machines built so far do not learn. They therefore cannot get better in *any* sense, let alone grow more intelligent by the millisecond. Third, what evidence is there that our own rate of intellectual progress is suboptimal? Evolution has tuned us to be excellent learning and thinking machines. It took four billion years to get us to this point, and I think we could probably hold our own if it came to a race, especially if we were the ones who designed the competitors.

Incidentally, many AI researchers are starting to place their faith in artificial evolution to do their work for them, which has perhaps created yet another source of unease for the general public. This is because it implies that we are letting our creations off the leash and will never be able to control them again. But I think that placing one's faith in evolution to design an artificial brain is often no more than an admission of defeat. We are saying that we cannot crack the problem ourselves, and if evolution cannot crack it either then that is not our fault. I am not criticizing all attempts to move in this direction – many of my friends are working on evolutionary approaches and they are very smart people. What troubles me about this line of attack is that many workers in the field fail to do their arithmetic. It took four billion years and a global population of millions of species and trillions of individual organisms to get to where we are today. In what sense do people think that they can do better than this, armed only with a digital computer and some wishful thinking? If you try doing the calculations, and are as optimistic as you can be about the speed of computers, the simplicity of the task and the cleverness of the researchers, you still end up with an experiment that may take a few million years to run. I do not have the patience to wait that long – what if the program crashes halfway through? So even if future intelligent systems are evolved instead of designed, any fears that they will then romp ahead and out-evolve human beings are simply unfounded.

Finally, some of these prophets of doom have been known to claim that AI is about to get itself into the worrying state in which machines take over the planet, largely because all the problems have now been solved, and all that remains for us to do is scale everything up a bit. This is absolute rubbish. We certainly do not know all the answers – we do not even know many of the questions. And scaling up what we already have will get people nowhere at all. Perhaps unsurprisingly, such claims tend to come from people whose roots lie in good old-fashioned AI – the sort that has just failed its own Turing test. Do not believe them!

So what will it be like, then?

That last paragraph makes it sound as if I think that AI is getting nowhere. That is not at all what believe: I am very optimistic indeed about the future for intelligence, both our ability to create it artificially and our capacity to understand and appreciate our own. Indeed, I recently got into trouble with some academics who accused me of being far too upbeat about the whole enterprise. The specific point of disagreement was a newspaper article in which I was quoted as saying that we might reach the sophistication of an artificial chimpanzee within forty years. They told me I should not go around claiming that AI will succeed in the foreseeable future because it gave the public the wrong message. Some reckoned it would take their lifetime just to reproduce the intelligence of a sea slug. For my part, I could not help feeling that if they were as pessimistic as they claim about their own chances of success, then perhaps they have chosen the wrong career. Just because the first fifty years of AI has gone so badly wrong, we cannot assume that the next fifty will go just as badly. Part of the reason for my writing this book is my belief that we have turned a corner, that our understanding of natural intelligent systems is no longer stymied by our inappropriate paradigms – and the future is starting to look much more rosy.

So, what will the future be like? Well, obviously I haven't a clue. I began this book with dire warnings about soothsayers, and I would be a fool to make predictions now. Nevertheless, I can tell you what kind of a future I am trying to work towards, and how that might feel for you and me if it happens.

The first thing worth pointing out is this: the near future for intelligent machines probably lies in artificial stupidity. That is to say, it is not androids with the intelligence of a human being that we should be expecting, but assorted machines, structures and appliances with the intelligence of a squirrel. I chose a squirrel as my example because I can see one now, through the window. But I could be talking about any mammal, up to and including many primates but definitely not including human beings. Squirrels are pretty smart animals. Imagine what it would be like if you could cut out the brain of a squirrel and mount it inside a different body, and then tweak its brain to give it a different set of motivations and goals.

The best analogy I can find is the Flintstones. Remember how all their technology used to be driven by living creatures? Owing to a slight mix-up with dates, these creatures were generally dinosaurs, and they 'inhabited' washing machines, playground toys and just about anything else mechanical. Strangely, the only thing that was not driven and controlled by a domesticated reptile was the motor car, which Fred and Barney had to move along with their own feet. The big difference between the Flintstones and the world I would like to see is that the dinosaurs in the Flintstones universally hated their jobs, while I expect synthetic life forms to enjoy every minute of their working lives.

Thousands of workers in AI are still trying to recreate the reasoning abilities of chess grandmasters. Many of my colleagues who work at the other end of the scale are trying to build systems with the intelligence of insects (or rather the intelligence of insect colonies – lots of dumb things acting in concert to create one smart thing, just like neurones). But very few people seem to be working on middling intelligence like that of a squirrel. Yet such a disembodied 'creature-on-a-chip' is an exciting possibility to contemplate.

I try not to think too much about actual applications when I am working, because I find it encourages the development of application-specific technology, when what I really want to do is create general-purpose solutions. I shall leave other people to dream up the applications – once they see what I have made for them I am sure they will think of uses for it. But imagine putting squirrel brains into, let us say, a set of traffic lights. Compared with the millions of sensory inputs and hundreds of muscles that a squirrel has to contend with, operating a set of traffic lights and talking to the traffic signals and other equipment nearby sounds pretty easy. Yet present-day traffic lights seem especially designed to snarl up traffic and cause gridlock. Trying to control a whole traffic network from a central control point appears to be futile. A single set of roadworks or a minor accident immediately changes all the rules, even supposing anyone can figure out what the rules ought to be in the first place. A bottom-up approach in which each signal is aware of the traffic flow nearby and the state of the neighbouring lights, and tries to make up its own mind about what to do, is much more likely to be robust. The problem is to work out what rules each traffic light should obey in such a decentralized system. But this is peanuts (or rather hazelnuts) compared with the challenges facing the average squirrel. If the mind of a rodent was placed into each signal, and the signals were rewarded for how well they managed to smooth the flow of traffic in their local area, then it seems plausible that it could work. In fact, a squirrel is rather 'overqualified' for this task, but if such general-purpose, self-learning machines were mass-produced and had many other applications (and enjoyed the tasks they were set), who cares?

Even really trivial devices such as light switches could benefit from the ability to learn by themselves, particularly if they were linked up to other devices such as clocks, video recorders (what wouldn't I give for a video recorder that programmed itself!) and toasters. In such a networked home, the ability of the individual devices to learn for themselves could be a great boon – far better than hoping that a human designer could cater for all the eventualities that might arise in a system with maybe hundreds of interacting parts, combined in different arrangements and used in very different ways by each owner.

Adapting automatically to the personality and habits of a particular owner is something that pets can do but present-day machines cannot. The engine management system for a car might be one application for such a facility. The creature would try to optimize the car's efficiency and second-guess its driver's demands for power. It would adapt to the particular driving style of its owner, and even to the road conditions.

I can think of various ways in which mildly intelligent, emotion-driven creatures could help people who are handicapped. If someone's brain does not connect properly to their legs, then perhaps their legs could be given a brain of their own. This embedded creature would like nothing better than to learn to walk, and would do so at its owner's command. Alternatively, imagine a creature buried in an electric wheelchair. Not everyone can exert precise enough control to handle a joystick. Some people can only blink an eye or twitch their chin. How can they gain control of a fiddly device like a wheelchair? It would be a great help if the wheelchair could learn for itself how to interpret its owner's desires. All the owner has to be able to do at a minimum is communicate to the wheelchair how well it is doing, which can be done by something as simple and natural as raising the eyebrows or frowning. Perhaps the ultimate disability aid would be a guide 'dog' for the blind. Real dogs do such a good job of this that it would be hard to better them, but perhaps an artificial guide would have its advantages, not the least being that it might be concealed in its owner's clothes.

Not all applications are quite so benign, I admit. I am currently working on a robot model glider, trying to give it the mind of an eagle so that it can learn to soar. I am doing this for scientific reasons, to test a specific theory, but it is clear that tiny autonomous aircraft such as this have military applications too. I do not think I am doing anything too scary, because most of the military roles for such devices seem to involve protecting life more than destroying it – anti-sniper systems for urban terrorism, or automatic air patrols watching out for trouble. Nobody would dare use a mechanism like this to steer a cruise missile. I do have to face up to the fact that some of the consequences of this kind of work may be ethically and morally difficult. But who cannot say that about their creations? Motor cars gave rise to tanks, the jet engine gave rise to the MIG as well as the Jumbo, safer explosives – thanks to Alfred Nobel – led to better ammunition as well as a prize for peace. All change has its dark side, but if the upside is better than the status quo, we owe it to ourselves to progress.

I can no more guess the future uses of living machines than Theodore Maiman, when he invented the laser, could have foreseen the development of the CD player (the most common application for lasers so far). Nevertheless, I am confident there are many, many ways in which the life can be put back into technology without requiring logical reasoning machines with the ability of a human being and a heart of stone.

The future has already begun

While we are on the general subject of how people will interact with intelligent machines in the future, it might be worth listing a few of the exploits of the people who already have some practical experience of artificial life forms – the owners of *Creatures*. If *Creatures* was going to have any influence at all over people's lives, the one thing I really wished for was that it might prompt questions in their minds about life and what it means. And it did this in bucketloads.

Take cruelty, for instance. One *Creatures* fan (who I believe is in the US Navy) started a Website entirely devoted to ways in which people could be cruel to their creatures. Four years later he is still running it. He devised various tortures to make their little lives a misery, and I think he did so with his tongue firmly in his cheek and a challenging grin on his face. I was so pleased about this (although I didn't dare say so publicly while I still represented the company that made *Creatures*, for fear that it would upset our customers), because it forced people to think about whether this really was cruel. I expected him to elicit some response from the other *Creatures* owners, but not quite such a hostile one as ensued. The poor guy received an enormous amount of hate mail, and was excluded from the *Creatures* Internet community for a long time. Much of his hate mail showed a greater regard for the creatures than it did for the life of this one human being. Less traumatically, I was pleased to see people start up Websites of their own in response, setting them up as rehab centres and adoption agencies to provide shelter for the poor victims of this virtual abuse. This was not the behaviour of people who felt that life had been debased by the existence of artificial beings, nor was it the behaviour of people who were scared of artificial intelligence. These people were going out of their way to protect and care for their creatures, and they held them in equal regard to other more natural forms of life. They may have been overreacting, but I was glad of the sentiment.

At the time of writing, there are a million *Creatures* owners out there, with untold numbers of pet creatures, both living and dead. There are around four hundred Websites devoted to the product, and the *Creatures* Web community is the biggest for any computer game, even outclassing the mighty Quake community. Most of these Websites are filled with fond stories of creatures' exploits, and have superfluous but still-loved creatures available for download and adoption. Many sites are devoted to understanding and tracking the genetics of creatures, including the conducting of patient and expert experiments on their evolution. Forty per cent of *Creatures* owners are female (much higher than the usual ratio for computer games) and, although most of them are in their teens, a significant fraction of the most expert and dedicated owners are adults in middle age.

So artificial intelligence is not all bad. Machines are not going to take over the world; they are going to become more useful, more adaptable and, above all, more friendly. People who have already experienced artificial life forms have welcomed them and learned to love them. I really do not know what the audience at the ICA were getting so worried about.

* *

CHAPTER FIFTEEN

VAPOURWARE

Bishop Berkeley destroyed this world in one volume octavo; and
nothing remained, after his time, but mind; which experienced a
similar fate from the hand of Mr Hume in 1739.

Sydney Smith, *Sketches of Moral Philosophy*

Deep down, perhaps what really frightens people about artificial life
and AI is not so much the machines running amok, but what the very
possibility of such things says about *us*, as human beings. If machines
can be alive and have thoughts and goals of their own, then presumably we need be no more than machines ourselves, and people simply
do not want to be told that they are nothing but clockwork. Yet I
started out by asserting a paradox that life is *more than just* clockwork,
even though it is *nothing but* clockwork. We have dealt fairly compre-
hensively with the latter idea: we examined the nature of the basic
mechanical processes that underlie all persistent phenomena, and we
looked in detail at how structures with at least a little personality, intelligence and autonomy can arise from them. Since we have now
reached the last chapter, I think we can indulge ourselves in a little
speculation about the first assertion as well – that life is more than just
clockwork. So let me risk delving into the three things that we human
beings hold most sacrosanct: our free will, our minds and our souls.

The prisoner in the mirror

When we look at future artificial life forms, we shall be holding up a
mirror to ourselves. If these creatures consistently behave like us and
appear to feel things the way we do, then presumably they *are* like us,

in essence at least. But they are just machines, and what is scary about this is that we know that they are automata, governed by immutable rules. No matter how independent they appear, they cannot really have free will, surely? And if they do not, do we?

I warn you now that I am going to argue against the existence of free will, at least in the relatively simplistic form that we normally view it. But be careful what inferences you draw from my initial arguments and please wait for my conclusion. Nothing upsets me more than black-and-white reasoning – just because I disagree with proposition A, it does not mean that I must necessarily be defending an exactly opposite proposition B. Most arguments have more than two sides and I usually find that I have a preference for a proposition C, where C is something that does not lie at either extreme, yet is not some wishy-washy liberal compromise. You should not assume that because I disbelieve in free will I must be an evil determinist. I am a determinist, but that might not mean quite what you think.

What I mean by free will in this context is our ability (as distinct from our political right) to choose our own actions and decide our own future. If we have such an ability, then our future actions cannot logically be determined by our past, or we would not be making a free choice, and this implies a breakdown of cause and effect. But what does that mean? Does it mean that certain effects have no causes?

An effect with no cause is surely a random one. Some people do seem to invoke randomness to help protect their idea of free will, for instance the supposedly random behaviour of objects at the quantum level, but this is really clutching at straws. Replacing grim determinism with pure arbitrariness doesn't do much to help us feel noble about our choices. In any case, 'randomness' is a poorly defined term and there may actually be no such thing. All sorts of silly, paradoxical and unhelpful conclusions are drawn about supposedly random events because there is a crucial distinction between two concepts that we can call *indeterminate* and *undetermined*, which people sometimes confuse or assume that the one implies the other. Just because something is indeterminate (by which I mean it cannot be known), we must not conclude that it is undetermined (has no prior cause). Randomness does not really mean 'un-caused', it simply means that an effect has causes that are either unknowable or irrelevant.

This might be because the relationship between cause and effect in a particular case is so hard to follow, even in principle, that the event is to all intents and purposes random. An example of this form of randomness is the fall of a coin. Whether the coin lands on its head or its tail is entirely determined by the way in which it was thrown, the disposition of the coin when it left the thrower's hand and the movements of the air molecules through which it falls. Each of these in turn has its own explicit causes, and so on ad infinitum. Unfortunately, as chaos theory demonstrates, the relationship between the initial conditions and the ultimate outcome is such that the tiniest inaccuracies in our knowledge of these initial conditions can amplify extremely quickly and make our expectation of the outcome hopelessly inaccurate. The way the coin lands is absolutely inevitable, given the initial conditions, but because the conditions cannot be known to an infinite degree of accuracy, the outcome is completely unknowable, and we can consider it to be a random event.

In another kind of randomness, the relationships between cause and effect are simply not relevant to the question in hand. An example of this kind of randomness is the level of chatter in a classroom. Every single speech act is caused by a mass of prior events, many (though not all) of which come from elsewhere inside the classroom. The moment that one child speaks is determined by when others have spoken and what they said. But in practical terms it is not the individual utterances that are relevant (to the teacher, anyway) but the general noise level, which can be understood in terms of probability. The chatter is therefore essentially a random factor in the eyes of the teacher, even though the causes are part of the system that matters to her – it is only the statistical qualities of this randomness that are important.

So invoking randomness cannot allay our fear that we are just deterministic clockwork, because even random events have definite causes. In any case, randomness (in the sense of causeless or externally caused events) is not even necessary to account for the richness of life. I prided myself when I wrote *Creatures* that I had used almost no random numbers in the program. Superficial things, such as making the creatures' eyes blink, were decided by a pseudo-random number generator*, and so were a few more important things such as the

---

* Computers are decidedly deterministic machines, and so cannot produce truly random numbers. But by invoking the mathematics of chaos they can produce sequences of numbers that have no discernible pattern to them. These are properly described as pseudo-random.

random mutation of genes. The user's interactions with the creatures could be viewed as random influences, but since the user would be responding to the creatures and the creatures to their owner, I count the user as a part of the system and not as a form of external random noise. Beyond that, the rich complexity of the creatures' behaviour is essentially entirely deterministic. Everything that happens has a collection of explicit causes and is absolutely inevitable, given the precise configuration of the system (including the user) at the start. However, it is totally unknowable and often quite surprising.

If there is no randomness in the fundamentalist sense of events without causes, then for the future not to be entirely determined by the past we must sometimes allow the laws of physics to be broken, or at least bent. Sometimes things might not follow the rules, so a cause could have different effects at different times. But at which times? If the rules change under specific conditions, then those conditions are just more rules (perhaps ones we do not yet know about) and everything is still perfectly inevitable. If they change whenever they feel like it, then we are back to randomness again. However hard you try, you cannot wriggle out of the fact that everything that happens in the universe is inevitable.

Inevitable, but not necessarily knowable. It might not give you much comfort to realize this, but it is important nonetheless. Take the modern fear and preoccupation with 'genetic determinism', for example. As we uncover more of the details of the human genome, we are beginning to understand some of the relationships between heredity and behaviour. This has led to much loose talk about 'the gene for homosexuality' or the 'gene for breast cancer'. This understandably scares or angers people because it implies that they are prisoners of their own ancestry. Whatever they do in life, they were clearly 'programmed to do it' by their genes. Yet to talk of a 'gene for criminality' is silly, and it is quite wrong to assume that genetic determinism makes us all into shallow puppets whose lives could have been foretold by looking at our chromosomes.

Genes do not code for behaviour, they code for proteins. Even the

---

Some computers do have specialized hardware to generate random numbers, usually by counting the decay of radioactive particles or the 'noise' made by electrons flowing through a diode. These are events whose causes lie outside the computer and so are random in terms of the system being considered. But it does not mean that the events are causeless.

genes we created for our creatures a little while ago code for physiological structures like chemoreceptors, not for explicit behaviour. Proteins do ultimately influence behaviour, but the route is extremely tortuous and a consequence of the whole assemblage of proteins and the environment into which they are put. Sometimes we can see the links between something genetic and something behavioural, but to say that the one causes the other is to make into a linear chain something that is really a complex and unknowable web.

For example, I am left-handed, and so is my mother. Left-handedness seems to run in families, so this trait may have been 'caused' by a specific gene or genes. Perhaps these genes did nothing more dramatic than twisting a particular protein one way instead of another, and as a result altered my development as an embryo. Left-handers have a very slight tendency to feel like outsiders in a generally right-handed world. We are just a little bit handicapped (or advantaged, if you are a tennis player or a fencer), and this perhaps gives us a mildly different viewpoint on the world than we might otherwise have had. There is also a tendency for the anatomy of our brains to differ from those of right-handers and I fancy (without any statistical evidence to back it up) that left-handers have a marginally greater inclination to think holistically. Perhaps this is because the two halves of our brains don't specialize quite as much as a right-hander's, which in turn might be caused by directional biases in the environment, such as the fact that (in Britain) we write from left to right regardless of our hand preference. So perhaps this book (being about holism, and perhaps also slightly left-field in outlook) is a consequence of the gene or genes that made me left-handed. But to say that I inherited the 'gene to write this book' would be absurd.

So even though our future is an inevitable consequence of our past, this does not make us mere puppets, acting out a pre-scripted drama. This is because our future, although not optional, is absolutely unknowable. And I say 'absolutely' unknowable with some confidence. Certainly we can predict some of the *general features* of our own and others' futures (indeed the ability to do this is exactly what intelligence is, and why it is so useful), but we have no hope of predicting it all in detail. To all intents and purposes, therefore, the future is yet to be determined. The vast bulk of the universe is highly chaotic in nature. We cannot simply predict the future by extrapolating from the past. Like any chaotic system, the future is a product of every single step of the past. The only way to know it is to act it out in its entirety, just as we would have to 'run the program in our head' to predict the existence of the glider in John Conway's Game of Life. If our future cannot be known, even in principle, we should not let ourselves get worked up by the fact that it is still inevitable.

In truth, we *must* believe we have free will – even that fact itself is inevitable. A human society that does not believe in free will cannot survive. It will not be a persistent phenomenon, and we know that non-persistent phenomena have no more than a momentary existence. For example, when someone commits a crime (as with any causal act) they are doing something that was absolutely inevitable in the circumstances. So in some sense we should maybe say they are not responsible for their actions. Yet if we do not punish them it is just as inevitable that crime will increase, and any society that does this will collapse and disappear. The concept of personal responsibility is therefore a crucial part of the feedback mechanism that makes our society (and even many of us as individuals) viable and self-maintaining.

I personally advocate something that I am starting to call 'cybernetic fatalism'. When somebody does what they actually do, we should take pains to recognize that they were 'drawn into an attractor' (recall how systems containing feedback tend to lock into certain states more readily than others) that we would very probably have been drawn into ourselves, if we were in their shoes. In fact, if we were in *exactly* the same circumstances (whatever that means) we would inevitably do the same as they did. Take wars, for example. Nobody starts a war, and probably nobody wants one, but wars still happen. My interpretation of history is that when a war is declared, the people who declared it found (or believed, which is much the same thing) that they had no choice in the matter. It is hard to find examples of people's behaviour where we cannot say (given a detailed knowledge of their situation and personality), 'Well, they would do that, wouldn't they?' We should therefore try very hard to put ourselves in others' shoes before we judge them for their actions. Nevertheless, judge them we must, because although we recognize them to be victims of circumstances, we must also believe them to be free agents.

At the same time we must realize that we too are slaves to our circumstances, but although our future is inevitable we must *believe* that we are responsible for how things pan out. In other words, we should always exercise our intelligence to the best of our ability, to predict where the attractors lie – where the slippery slopes of life are leading us. Like the ridge-runner in Chapter 8, we must try to avoid getting into situations from which we cannot escape. It is our responsibility to try not to become victims of circumstance – to avoid getting too far from the ridge. When justice is called for, we should be held to account for the degree to which we could have predicted disaster and yet did not, or did not act to avert it.

I think there is a lesson to be learned here, and a crucial balance to be sought. There must be a proposition C that lies somewhere between proposition A – we are all free agents, and so whenever we do wrong we must be held liable for our actions – and proposition B – we are all slaves to our circumstances, and nobody can be held responsible for anything. Our society will be optimally self-sustaining when it finds the right mix (or rather the right logic for deciding) between compassion and justice. If we do not find the right balance, the pattern of society will break down; if we do, we shall be a part of a stable persistent phenomenon for a little longer. Whether or not we, as human beings in the early twenty-first century, shall find that right balance is both inevitable and unknowable. But we must *believe* that we are seeking it, because only systems that seek a balance can ever find it.

Nobody in this universe or outside it can know what the future holds, even though it is already set in stone by the mere fact that one thing invariably follows another. So when I sit on the seashore and discover my thoughts beating in synchrony with the lapping waves; when I see that my past and my future are laid out among the ebbs and flows of cause and effect, I am happy to be a part of that great dance – to play my role and hope that I will do it well. The things that I cannot ever know and that no one can ever tell me are a mystery yet to be revealed, and that is enough to make me free.

Mind over matter

And so to the second great taboo – the human mind. Actually, 'human' is probably not the right boundary to use. Some other animals have minds, too. A few biologists still doubt this (perhaps to make themselves feel better about lopping the heads off their experimental subjects), but many of their own arguments work equally well against themselves. Whatever reason someone may have for doubting that dogs have minds, we can often use the same logic to assert that he or she does not have a mind either. Perhaps the only time when this is not true is when we are dealing with anatomy. If you say that a particular animal has no mind because its brain lacks feature X, I may disagree with you about the importance of this feature, but at least I cannot claim that you have no mind either, because presumably you checked in advance that your brain does have feature X.

So what is feature X, and what is a mind anyway? I can offer some clues to the first part of this question, but the second part is rather more difficult. Another word used in connection with the mind is 'consciousness', or sometimes 'self-awareness'. These are horribly slippery terms. Is a thermostat conscious because it is 'aware' of its environment (the temperature in the room)? Those who say that it is are debasing the term so much that it is no longer useful, and we would then need a new term to describe what we have, which seems qualitatively rather different. I do not have space here to discuss the nature of phenomenal consciousness and self-awareness, but there are plenty of books on the subject, if rather fewer conclusions. Let me start by assuming that you already intuitively have some sense of what a mind is because you own one – or, more accurately, because you *are* one. Then we can go on to look for factor X and see if anything useful can be drawn from it.

One thing I should make clear here is that I do not believe the creatures I have described to you in these pages are conscious. They are alive, but they do not have minds. The biologist Richard Dawkins very kindly described my creatures as 'quasi-conscious', and I think this is a really nice term for what they are. As systems, they are the recipients of information about their environment (they are 'aware' of the state of the world). They also have information about their own internal state and their actions, so they are 'self-aware' too. But if you knock on their door, I think you'll find there is nobody home.

I believe that this is because they are trapped inside a deterministic world in a much less subtle way than we are. This is because, like insects and starfish and many other creatures, they are locked into a sensory-motor loop. The environment tells their senses what to do, their senses tell their brains what to do, their brains command their muscles and the resulting actions alter the environment, which in turn alters their senses once more. Nevertheless, at some point (or points) in animal evolution, something special happened which makes humans and some of our animal cousins different from the others. That something is the capacity to imagine.

You and I do not live in the real world at all: we live in a virtual world inside our heads. Most of the time this dream-world is closely synchronized to external reality, but sometimes it is free to live out a different future of its own. There is plenty of evidence to show that we are not really conscious of how things *are*, but only of how we think they are, in our own models of the world. When I trained as a teacher the most important thing I learned from watching children is that their brains are fully competent from an early age – it is only their models of the world that need a bit of tinkering with. The program is all there, but it lacks good data. Watch any eight-year-old girl (it is more noticeable with girls) and you see a grown-up woman. It is just that this woman is entirely capable of believing in fairies and often confuses fantasy with reality. This is because her reality is a fantasy too – her mental model of the world is not yet finely tuned to the real thing.

It is difficult to say when, why and how this faculty of imagination arose, but I think I can see several stages in its evolution, perhaps starting with the ability to distinguish self from non-self. When we feel a pressure on our bodies we need to know whether we have bumped into something or something else has bumped into us. For that we need a sense of where our body is in space and how it is arranged. This sense is known as 'proprioception' in its simpler forms and 'body image' in its various higher forms. Similarly, if we hear a sound we need to know whether it comes from outside or is the sound of our own voice. Sometimes this system can go wrong. Perhaps all of us, when we are extremely tired, have experienced the feeling that our own voices are echoing back at us. For similar reasons, people who have lost limbs often still feel pain in them, and sometimes they even believe they can still move them. Here we see some clues to the basic biological mechanisms that underlie our ability to keep track of our own bodily and sensory state as a kind of model of ourselves.

Perhaps a closely related system is the one that allows us to predict the future position of our muscles. The signals from the stretch receptors in muscles often do not reach the brain quickly enough to take part in a simple feedback loop. If they did, we could keep on flexing a muscle until these sensors told us it had reached the position we intended, and such a feedback mechanism (also known as a servomechanism) would cope automatically with different degrees of resistance to movement: for example, it would still reach the right position when we were carrying a heavy weight. Often our muscles do work this way – when we balance on a log we are 'servo-ing' our body to maintain an upright posture. But since we often move quite quickly and these signals take a finite time to arrive, there is a risk that our limbs will frequently overshoot their target (an effect called 'hunting', which is similar to the 'ringing' in negative feedback loops that lack damping). So instead of relying on knowing where our limbs are, we must have a system that can *predict* how much energy to put into a muscle and where the limb will end up as a result.

From such neural models of ourselves, a more general ability to predict the future and make plans may have arisen. Many of the plans we make and carry out are entirely subconscious. We may need to go to a different room, and this requires us to plan a route, think about opening doors and suchlike. To execute the plan we must map out smaller sequences of muscle movements and respond to discrepancies between the overall plan and reality (if we miss our footing, say). By adulthood we can usually carry out all of this subconsciously, and our minds simply have to set the goal, but at an earlier stage in our lives – the first time we experienced walking, or navigating, or going to the shops, or whatever – we had to be consciously aware and make deliberate plans every step of the way. These skills only became unconscious after some part of our brain learned to copy what our consciousness had been doing.

One high-level consequence of such mental models is the ability to rehearse things in our heads. Not only can we predict what will happen if we continue as we are, but we can also predict what might happen if we were to take any of several different courses of action. If we have something especially nerve-wracking to do (a public appearance, say), we can play out in our heads how it might go in reality, and make contingency plans about what to do if things go wrong. Only some creatures can do this. Humans are remarkably good at it, to the extent that we can invent stories and tell them to others, and the listeners can build the story-world inside their own heads and play it through for themselves.

Out of all this comes our sense of self. The neurology of such mental modelling systems is the subject of my current research into artificial life. I am trying to build machines with imagination – not just passive internal representations of the outside world (this has already been done to a degree), but representations that inevitably and automatically lead to the sequencing of movement, the production of plans and, ultimately, the ability to 'direct operations' in the way that happens when we are conscious. At the moment I have some exciting hunches about how the lower parts of such systems might be constructed, and I am trying to make a machine that can build mental models and use them to coordinate its actions. So far I have absolutely no idea what mechanisms give rise to the highest level of volition, conscious thought, but perhaps by building the next layer down I shall gain some insights.

I wish I could tell you what a mind is, and how to construct one, but as yet I cannot. I can only give you some clues about where such a phenomenon may have come from. There is one observation worth pointing out, though: the mind seems to me to be an emergent, persistent phenomenon that arises from inside a *virtual* space (the world of our imagination). We seem almost to be products of our own fevered imaginations. This idea seems hauntingly familiar – in the right hands the digital computer is an imagination machine too. It can be made to 'imagine' (simulate) a space and some virtual components that we can plug together to build something alive and intelligent. That creature is a machine made out of imaginary parts and it inhabits cyberspace instead of real space. If our minds are virtual machines made out of imaginary parts conjured up by the biology of our brains, then presumably an *artificial* consciousness will be a persistent phenomenon conjured up by a simulation, produced in a virtual machine which itself is conjured up by a simulation that exists in the software of a digital computer. I am sure there are some important lessons to be drawn from this, but as yet I am unable to see what they are.

Product of a simulation or not, the mind is still a thing, no less real than an atom. If neuroscientists or physicists or anyone else tells you that consciousness is just an illusion, don't listen to them. Illusions look like something they are not, but the mind is exactly what it seems: it is something more than the sum of its parts.

Posterity

Life may be a persistent phenomenon, but one thing we are sure of (and frequently terrified by) is the fact that individual examples of it do not persist for ever. Creatures are eaten, have accidents or fall sick and die. The pattern is disrupted and cannot reassemble itself. The system swerves off the ridge and is never seen again. Each of us knows that the persistent phenomenon that is our body will eventually crumble and decay, but whatever our religious persuasion, most of us cling to a belief – or at least a desperate hope – that the same fate will not befall our minds. Every night we go to sleep and for a few hours we lose consciousness, but it always manages to reassemble itself the next morning. The memories imprinted in our brains tell us who we were the day before, and so our existence appears to continue seamlessly.

Our minds are products of our bodies, but they are emergent, persistent phenomena in their own right. So do they have any independence from their substrate? Can there be such a thing as a disembodied mind – do we have a soul? Certainly we are familiar with the idea that the same pattern in time and space can exist in many forms and be transferred from one form to the other and back again. Every time we make a tape recording we are converting a pattern in time into a pattern in space: from sound waves to magnetic stripes. Later, the original pattern can be reconstructed and we can barely tell the difference. Similarly, a wide variety of information sources can be stored on a computer disk and then reconstructed. The notable difference here is that the pattern on a computer disk appears to bear little relation to the pattern of flowing electrical signals that gave rise to it, unlike the audio tape. Computer disks can also store *potential* information, in the sense that a computer program can encode rules as well as facts.

But this kind of immortality is pretty dull, unless you hope that, by recording the state of your brain, you can either pass into a new machine and live on or be reconstructed at some later stage. I think both of these things are theoretically possible, but in practice I think it is very unlikely to happen before you and I shuffle off this mortal coil, if ever. Indeed, there is something disconcerting about the very idea, because (thanks to the first law of cyberspace) you cannot *move* into a new host – you have to be copied there. The original you would still have to face death, and the new you might only be one of many, because copying information is essentially free. Some people speculate on the possibility of dismantling your biological brain, cell by cell, and replacing it incrementally with new silicon neurones. Many explorations of these ideas fall foul of poor reasoning and a failure of the imagination, but there is no reason in principle why you should not convert yourself piece by piece into some other technology. After all, our bodies are replacing bits of themselves all the time. Whether the result will still be you and whether the original you is now dead are rather more difficult questions.

Most of us cannot wait around hoping for a techno-fix, and are more interested in the question of whether the mind can still be a mind without any kind of a body at all – do we have a soul that is free to wander? Certainly it is a feature of some kinds of self-organizing system that they seem to have a tenacity that goes beyond their original instantiation. Even our bodies, as described in Chapter 2, are patterns that persist despite their component parts being in constant flux. Whatever you are, you are not the stuff of which you are made. Matter flows from place to place and momentarily comes together to be you. In fact, matter flows from place to place and becomes our children too – in our neck of the universe at least, more matter is now life-shaped than ever before in its history. This persistence of form is more the rule than the exception, and it was observations like this that I hope have encouraged you to conceive of these patterns as being things in their own right.

Although the *rules* that govern a phenomenon at one level of description can be described in the language of the level below, we certainly cannot infer from these rules the limits of the higher-order construct's *behaviour*. Human societies are ultimately a product of Newton's laws of motion, and these impose certain limitations on any physical system. But they do not really have much bearing on what societies can and cannot do. So who knows? Simple emergent phenomena do not seem to have quite enough 'oomph' to get up and walk away from the substrate that gave rise to them, but I do not think we can necessarily extrapolate from that to minds, which are easily the most sublime of persistent phenomena that I know of (bearing in mind that I may be quite unable to conceive of anything more sublime).

I wish I could say more, but I cannot. I have been studying persistent phenomena for years, with the feeling that there is something really important right in front of my nose and I cannot see it. Maybe one day I will trip over it and recognize the truth. Or maybe one day this particular persistent phenomenon will cease to persist any longer, and the mystery will cease to matter to me.

I shall leave you with one intriguing thought about souls. I have not been able to explain things as clearly as I would like, but I hope you can see that higher-order phenomena (such as minds) are often constructs of what is, in effect, a simulation. My sweetly dim little creatures are real objects in their own right, but their constituent parts are really just a sham, simulated on a computer. My own mind seems to be intimately wrapped up in a similar way, inside the simulation machine that allows my brain to imagine things. The human brain is the most awe-inspiring and powerful simulator in the world – the depths of our imagination have never been plumbed and perhaps never will be. An important part of the duties of our imagination is to conjure up simulations of other people so that we can predict how they will react when we interact with them. Humans are far and away the best animals on this planet at creating models of their fellows, and we do it with consummate ease. Our models are sophisticated and emergent, rather than shallow and explicit – we are truly able to place ourselves in other people's shoes. So if I am a product of my brain's imagination, can I be a product of yours too? If you imagine me (assuming you know me well enough) and I appear inside your head and start behaving as I would in real life, then have you conjured up a genie? Is the 'me' of your imagination just an empty shell – something that behaves *as if* it were alive – or is there really a 'me', however reduced and impotent, momentarily living inside the virtual world of your mind?

Most of us feel driven to erect memorials to ourselves, to leave some lasting impression on the world through which to be remembered in other people's minds. We know that our original instances will fade and die, but perhaps feel that clones of ourselves may live on in the imagination of others. When we remember the dead, do we literally re-member them, and bring them back to life? Who knows, but we are certainly driven not to forget or be forgotten. To that end, I would like to dedicate this closing chapter to the memories of the first creatures I created, who taught me so much about life, the universe and everything. May their future descendants find a place in their hearts for the memory of Ron and Eve.

BIBLIOGRAPHY

* *

The Matter Myth, Paul Davies and John Gribbin (Viking, 1991)   An anti-materialist view of nature as seen from a physicist's perspective. Particularly interesting for its discussion of solitons – propagating, self-maintaining persistent phenomena that may help to explain how the universe bends itself up to form particles.

Does God Play Dice? The New Mathematics of Chaos, Ian Stewart (Blackwell, 1989)   Everyone should have heard of chaos theory by now, but if you haven't then this is the book to tell you about it.

The Planiverse: Computer Contact with a Two-dimensional World, A.K. Dewdney (McClelland & Stewart, 1984)   This story crystallized my amateur interests into a focused desire to create virtual worlds filled with artificial life forms. It tells the fictional tale of a student project on computer simulation that somehow became synchronized with a real, two-dimensional world of bizarre creatures. Sadly now out of print.

The Selfish Gene, Richard Dawkins (Oxford University Press, 1976)   Of course you will already have read this. It is the definitive guide to survival machines – to persistent phenomena of the genetic kind.

Frankenstein, Mary Shelley (Quality Paperback Book Club edition, 1994; first published in 1818)   I include this largely because Frankenstein and his 'monster' deserve more sympathy than they usually receive – the common impression of Frankenstein owes more to films than it does to Mary Shelley.

Alan Turing: The Enigma, Andrew Hodges (Burnett Books, 1983)   An enthralling and sensitive account of the father of the digital computer and one of the first people seriously to consider ways in which to make a machine with a mind of its own.

Out of Control, Kevin Kelly (Perseus Books, 1994)   The bible of bottom-up. A wide-ranging review of new ways of thinking about the natural and artificial worlds.

Artificial Life: The Quest for a New Creation, Steven Levy (Pantheon, 1992)   A biographical story of the granddaddies of A-life. A great starting point, and a place to meet many of the characters who did the seminal work in this field.

Complexity: The Emerging Science at the Edge of Order and Chaos, M. Mitchell Waldrop (Simon & Schuster, 1992)   Another biographical introduction, this time to complexity theory. Many of the characters in Levy's book on A-life also turn up here, which shows how closely related the two fields are.

The Emperor's New Mind: Concerning Computers, Minds and the Laws of Physics, Roger Penrose (Oxford University Press, 1989)   Since I mention Roger Penrose critically a couple of times, I include this book in the list so that you can decide for yourself. Even though I disagree with his conclusions, it is an excellent book that romps through a huge range of ideas.

Phantoms in the Brain, Vilayanur Ramachandran *et al.* (Fourth Estate, 1998) A good source of anecdotes about what goes wrong when different parts of the brain fail. Especially interesting for its reports of amputees with phantom limbs, which provide some insights into the biological substrate of the imagination.

At Home in the Universe: The Search for Laws of Complexity, Stuart Kauffman (Viking, 1995)   A study of the emergence of order and the origins of life. Includes accessible yet detailed information on the behaviour of autocatalytic networks.

http://www.creatures.co.uk   This is the main Website for the *Creatures* community, provided by the product's developer, Creature Labs (Cyberlife Technology Ltd). As well as reading about the activities of *Creatures* owners here, you will find information on the range of *Creatures* products.

http://www.cyberlife-research.com   My own company's website. Feel free to drop in and see how my latest research is progressing.

INDEX

* *